Into the Field of Suffering

Into the Field of Suffering

Finding the Other Side of Burnout

DAVID SCHENCK WITH SCOTT NEELY

OXFORD
UNIVERSITY PRESS

Oxford University Press is a department of the University of Oxford. It furthers the University's objective of excellence in research, scholarship, and education by publishing worldwide. Oxford is a registered trade mark of Oxford University Press in the UK and certain other countries.

Published in the United States of America by Oxford University Press
198 Madison Avenue, New York, NY 10016, United States of America.

CIP data is on file at the Library of Congress

ISBN 978–0–19–766673–9

DOI: 10.1093/oso/9780197666739.001.0001

Printed by Integrated Books International, United States of America

For

Larry Churchill

"A character justly proportioned in graciousness and honor."

Marcus Aurelius praising his teacher, Claudius Maximus

—The Meditations

He knew that the tale he had to tell . . . could be only the record of what had had to be done, and . . . would have to be done again in the never ending fight against terror and its relentless onslaughts, despite their personal afflictions, by all who, while unable to be saints but refusing to bow down to pestilences, strive their utmost to be healers.

—Albert Camus, *The Plague* (1991, p. 308)

Contents

Acknowledgments

We thank those who made the composition and completion of this book possible:

- First off, Lucy Randall, our extraordinary editor at Oxford University Press. She was a champion of this book from the very beginning, providing all the perceptive guidance and technical assistance one could hope for from one's editor. She kept the lead author on track, encouraging him at his worst moments. No Lucy, no book. It's that simple.
- For careful readings of the many drafts of the manuscript, and trenchant comments on every part of them: Larry Churchill, Art Frank, Wylie Burke, Helen Chapple, Keith Hagan, Kathy Meacham, Elizabeth Mack, Frank Harris, Jim Abbott, and Jane Perrin.
- For double-checking the presentation of clinical detail and for our ongoing friendship: Shelly Ozark.
- For their careful work and detailed suggestions: the three readers who reviewed the manuscript for Oxford University Press.
- For capable assistance in the production process: Paloma Escovedo.
- For extraordinary care and insightfulness in preparing the manuscript, resulting in a much-improved book: Katie Haywood.

Scott Neely thanks the nurses, doctors, administrators, chaplains, therapists, and pastors—healers, every one—who have shared with me your vocations and your struggles to keep going over the last three years—who have each found a way to break through. You have held the world together and shown us all a more excellent way.

David Schenck thanks the following:

- My colleagues at Vanderbilt University Medical School and the Medical University of South Carolina (MUSC), who consistently supported this work on burnout and moral distress, and made available opportunities for presentations and workshops, where early versions of these concepts and approaches were tested out. At MUSC, I would single out Danielle Scheurer—best boss ever. Good wishes to everyone.
- For inviting me to facilitate with him meetings for his physician circle, Courage to Heal, where many of the concepts and exercises offered in this book were developed—my fellow seeker, friend, and mentor: Keith Hagan.
- For collaboration on my first burnout workshop for nurses, and ongoing support and friendship ever since: Allison Bannon.
- For dozens of conversations about her professional development and mine, reliable advice on workshop structure, and a continually unfolding friendship: Jess Bullington.
- For teachings essential to my life and my work: Cheri Huber and Brugh Joy.
- For ongoing support of all kinds—spiritual, financial, and existential: my marvelous sisters, Dee and Patti.

Scott Neely, friend, colleague, and co-author—intellectual and artistic collaborator across three decades. Our work together has pushed me steadily into deeper understandings of what this book is about. My best student, who now teaches me—what a blessing!

And, finally, in the place of honor, my friend, my colleague, my counselor, Larry Churchill. What has he not given me? A whole new career in the sixth decade of my life, and joint authorship of two books and two much-discussed articles. Ours is a friendship that has shaped and altered the course of my life at critical junctures time and time again. Without his encouragement, enthusiasm, editorial advice, and advocacy, this book would never have escaped the bonfire. More concretely, he read and commented on countless versions, drafts, revisions, and schemes. We had dozens of phone calls and exchanged endless streams of emails. No one could have been more available, or more generous. All possible respect and gratitude, my friend. This book is dedicated to you.

A Note on Sources and Method

The ideas in this book were developed primarily in the context of healthcare in the United States, and most especially in hospice settings, free clinics, and adult and pediatric intensive care units. But much that lies behind the book comes also from work with people in the community agencies and organizations in some of our poorest and most marginalized communities in the Southeast—from coal-mining counties to sugarcane country to mill villages—from the homeless, those stricken in the AIDS epidemic, and communities displaced by floods.

Sources

The cases described in this book are drawn almost entirely from David Schenck's prior research and practical experience. Specifically,

- Fifty interviews with clinicians for a study of clinician–patient relationships, published as *Healers: Extraordinary Clinicians at Work* (Schenck and Churchill 2011); interviewees included physicians from a variety of specialties (surgery, cardiology, pulmonology, internal medicine, family medicine, and psychiatry), nurse practitioners, a nurse midwife, and a variety of complementary and alternative practitioners (including acupuncturists, chiropractors, massage therapists, and a craniosacral practitioner).

- Fifty-five interviews with patients of the above practitioners, for a follow-up study of clinician–patient relationships, published as *What Patients Teach: The Everyday Ethics of Health Care* (Churchill, Fanning, and Schenck 2013).
- Dozens of workshops, presentations, and lectures on healing relationships in medicine and moral distress, with anywhere from a half-dozen to 120 participants, across nearly two decades.
- Ethics consultations with a variety of medical teams in two major medical centers over a 10-year period. Conservatively speaking, more than 100 cases a year, ranging from the very simple to the quite complex. In addition to those formal consults, countless "curbside consultations" and individual debriefings.
    - Of those formal consultations, roughly 15 percent involved family conferences in adult and pediatric intensive care units.
    - Of those formal consultations, roughly 10 percent involved formal staff debriefings with physicians, nurses, and physician residents, focused on complex cases with challenging moral, legal, and clinical components.
- More than 30 years volunteering for a variety of hospice organizations, doing direct patient and family care.
- Six years coaching executive directors of small, community-oriented nonprofit organizations that were focused on healthcare, housing, racial justice, and economic development.
- Twenty years of teaching philosophy and religious studies at the undergraduate level, with an established research record in religious ethics, philosophy of religion, and phenomenology.

Names of clinicians and patients have been changed in all cases recounted. Details have been altered or omitted to preserve privacy in all patient cases, and in many clinician cases. In certain illustrative scenarios, details from two or more cases have been combined.

Method

The material constituting this book was first presented and refined by David Schenck at workshops and conferences over an eight-year period. In summer 2019, Schenck and Scott Neely recorded approximately 15 hours of conversation in nine separate sessions. These conversations were guided by the outlines and PowerPoints from Schenck's workshops and conferences. These recordings were professionally transcribed, and an editing process was begun.

Schenck and Neely decided, for the sake of compression and clarity, to compose solo voice chapters, with Schenck as the author, to make up Part I. Dialogue selections with Schenck and Neely, intended to supplement and clarify key themes, constitute Part II. Although the core material was developed by Schenck, every sentence in the manuscript has been shaped by conversations with Neely, as well as by his comments and suggestions on successive drafts.

Finally, the reader will find that we press constantly against the normative rules of sentence structure and punctuation. This is done to maintain rhythms of conversation and to approximate the direct address that characterizes workshops and debriefing sessions.

Invitation

For You Who Do This Work

This book is written for people who spend the bulk of their days working with others who are suffering. What I want to offer here is an invitation to conversation: a conversation with me, a conversation within yourself, and a conversation with the people with whom you work. A conversation about what it means to spend your days with people whose lives are disrupted by illness or overwhelmed by brokenness. That can be brokenness in bodies. It can be anything from flu to cancer, from sprains to AIDS. It can be traumatic brain injury, severe mental illness, or debilitating chronic conditions. Or it could be families in neonatal units dealing with children with debilitating genetic defects, struggling to make decisions about if, when, and how to end lives. Or violence on the street, or the devastation of a flood-ravaged community.

What is it that we can learn from these situations? I put an emphasis on learning because I do believe that if we don't learn every day, if we don't learn every hour, this work will destroy us. It's important to use a word as strong as "destroy." There are reasons not everyone does this work, and there are limits to how much of this work anyone can do.

What we must keep asking ourselves is: Why do we do this work? Why are we drawn to those who suffer? What is it that's good about this? And what not so good? What is it that is admirable, and something that we should show enormous compassion and respect to ourselves for? And what is there in it that is self-destructive, that

Into the Field of Suffering. David Schenck with Scott Neely, Oxford University Press.
 DOI: 10.1093/oso/9780197666739.003.0001

we should be constantly questioning and challenging? To answer these questions, we must get past the assumption that attending to the suffering is entirely saintly and noble, or that it is a symptom of a deep, perverse drive.

Much attention is rightly being focused on people who serve on healthcare's front lines. The framework for these discussions tends most often to be the terms "burnout" and "moral distress." These terms certainly point to very real experiences, very painful experiences. Unfortunately, they also often carry with them judgmental messages: "You are failing." "You must do better." What I want to say to you here is that, while these terms point to realities for those of us who have worked with suffering day in and day out, they may not help us move and grow into the deeper ranges of compassion and recognition and attention that are possible, unless we supplement them with other constructs and insights.

When I say that I want to encourage a conversation within you, part of what I have in mind is this: We are all multiple selves. And one of the things that happens when we work with people and communities that are suffering and in pain—physical pain, psychological pain, spiritual pain—is that distinct pieces of ourselves often respond quite differently to what is going on in front of us. Many different feelings are stirred. We may be terrified, and at the same time move with great compassion, while being also completely exhausted, and madly energized.

This configuration of what, in ordinary life, would be experienced as contradictory, this discord of feelings all at once, is commonplace for those who work in intensive care units (ICUs) or in our most distressed communities. I know this by listening to the people who are there, by watching them do their amazing work. And I know this by having been there myself in these communities, on these streets, in these units, in the rooms with people who are dying.

What you will not find here is a summary of the literature in professional journals. What I offer instead is something much more direct. I want to offer to you—whether you are a nurse in an ICU, a physician in an emergency department, a worker on the street attempting to prevent daily violence, a staff member in an outpatient clinic or medical practice—a voice giving you permission to be in pain. Giving you permission to grow. Giving you permission to learn. Giving you permission to be enraged. Giving you permission finally to be kind to yourself. And kind to the world.

Trained in world religions and comparative literature as I was, I was familiar with various depictions of paths of life transformation. And I knew that these narratives tended to have common features, including elements of call and initiation, periods of activism and of darkness, moments of illumination. I can't say exactly when I first realized that the hospice workers, the ICU nurses, and the community agency staff I was working with day in and day out were themselves on paths of transformation. But I gradually began to recognize classic patterns in the stories my burnt-out and morally distressed colleagues were telling me. There was a friend who, like Dante, was lost in a dark wood and beset by beasts. Here were colleagues behaving like Joseph Campbell's storied heroes and heroines, meeting up with their monsters. Here were medical students and residents going through initiation rites similar in structure to those described by Arnold van Gennep and Victor Turner. There an ICU attending shaped by experiences like those analyzed in Mircea Eliade's work on shamanism, and there a neurosurgeon enduring John of the Cross' dark night of the soul. All around me my academic training was coming alive in people serving in the field of human suffering.

Through thousands of conversations and hundreds of interviews, working with participants in dozens of workshops over nearly two decades, all the while consulting in the nonprofit sector and academic medical centers, I kept seeing the basic patterns of inner and outer transformation that countless spiritual seekers across time and space have lived out. I was seeing bodhisattvas and saints in training, in formation, at work. Yet these very same people did not know they were on this path, did not fully know how amazing they were and, for the most part, did not have access to the wisdom of spiritual teachings that have traditionally guided such development. People working in healthcare and social services, and people working in or studying communities shaped by spiritual disciplines are, in my experience, mostly not much aware of each other—much less in communication. In fact, these communities are unfortunately often hostile to each other, due as much to the dogmatism of medicine as to the dogmatism of religion.

Despite this, I gradually found myself drawing on my academic training in religious studies, and my personal training in meditation, to understand the experiences I was hearing about that were more commonly spoken of as moral distress and burnout. During this period, I was also struggling with my own burnout recovery, which involved drawing on those same resources, as well as psychotherapy and trauma-release bodywork.

The book you have in your hands is woven of these three strands: (1) the experiences of moral distress and burnout of people courageously facing suffering day in and day out, (2) the study of spiritual narratives and practices of transformation, and (3) my own plunge into burnout. Behind this book lie years of intense personal exploration. Behind it lie years of leading workshops and counseling individuals struggling with overwhelming suffering, while working without adequate institutional support, and the anguish and depletion that result from that. And all that building on decades of volunteer work and trainings in various hospice settings. Gradually I found ways to introduce elements from traditional practices of transformation into the workshops. I also found, as many teachers have across the centuries and across cultures, that at certain moments, narrating elements of my own struggles and my own transformation was useful, and so I share certain pieces of my own story here, too.

While I do draw on contemporary research on the topics of burnout and moral distress, these chapters and dialogues are rooted primarily in the medical humanities. I fear that if we go only with the sciences—social and biomedical—as resources for addressing moral distress and burnout, we will have missed the most essential thing of all: the address of one human heart to another. That address—heart to heart—is what I hope to achieve here.

Everyone who works with patients and clients develops ways of understanding how healing happens, and what their role in this healing can be. Caregivers participate daily in the stories and processes of healing and recovery of their patients. Almost everyone who works with patients is aware that this business of healing and recovery can also be a process of transformation and rebirth. What I will take pains to argue as we progress is that we as healers, we who stand with those who suffer and suffer intensely, are ourselves of necessity on a similar path of transformation by virtue of standing in the field of suffering offered to us by our patients and clients. By exploring these dynamics, we can, along the way, come into new understandings of moral distress and burnout, and of how to address them in our lives.

I envision two primary audiences for this book:

1. Those in healthcare who know themselves to be struggling with moral distress and burnout
2. Those who work with intense, at times extreme, suffering in ICUs, emergency rooms, hospices, and safety-net community agencies

People who are curious or concerned about those working in such settings—partners, families, neighbors, colleagues, or other readers who have seen headlines about COVID-19 ICUs over the past two years—may find here insight into what these caregivers have gone through. Administrators, suppliers, and a host of others whose work touches on the overwhelming challenges frontline workers face will also find valuable the perspective introduced here.

In addition, lessons derived from examining moral distress and burnout in healthcare settings may be of use to people struggling with these career-disrupting trials in any sector.

Finally, I hope this book will benefit anyone who wants to understand themselves and their own lives better, as intense and traumatic experiences are intrinsic to the human condition.

The text is written to be accessible to the general reader. There is no technical terminology, and the cases drawn from the arena of medicine are written in layperson's terms, not those of clinicians. While it is true that the bulk of the vignettes and illustrative examples are drawn primarily from ICUs, hospices, and frontline community agencies, the lessons learned there can be applied across the board in healthcare—and indeed in any depleting workplace. Burnout is everywhere in our unhealthy culture.

There has been much discussion of the causes of burnout in the past few years. The burdens and impossible stresses endured by healthcare providers during the COVID-19 pandemic have intensified this discussion. There are those who maintain that burnout and moral distress are primarily, if not solely, caused by oppressive and unresponsive systems and institutions. Some analysts may feel that approaches like the one in this book, which focus on individual development, are misguided and, worse, serve to undermine efforts to overhaul healthcare itself.

Yet, regardless of what causes burnout, it is utterly essential for each of us to figure out what to do as we feel it approaching, as we go through it, as we come out on the other side. We have the power to act, to claim our evolving vocation once again. This will involve two pieces:

1. Advocating for institutional change and seeking to improve conditions in the unit or office where you work, the organization that employs you, and the healthcare system itself. But this will surely be a long, slow, arduous process—and there

are few clear plans about how to accomplish this in the current political and economic climate.

2. The other work is more personal. It is focused on the individual clinician or practitioner or agency staff member. How do you survive and grow as you work within these deeply flawed institutions? As you go through burnout and deal with moral distress?

This book addresses institutional vectors causing burnout and strategies for engaging them in several of its sections. But the primary emphasis is on personal transformation. Certainly, much more needs to be written about the healthcare system and burnout. But that is not the task we are undertaking. We are focusing on the capacity within each of us to pass through burnout into renewed vocations. This is the conversation I invite you to join.

It is important that you realize that I do not initiate this conversation out of a sense of having gotten over burnout, or past moral distress. These are issues I struggle with. One of the problems with the casual use of the terms "moral distress" and "burnout" is that it suggests one shouldn't be having these kinds of problems. Sometimes this goes with advice not to get enmeshed with patients, to get over ourselves, to toughen up. These messages are not helpful.

This is a book of exploration. It is not a guru book. There is no monopoly of wisdom here. There is, rather, a path: What is it that we can learn from being with people who suffer? And especially, how can we learn to be grateful to them? Because what is happening, when we serve in these profound ways, is that the people we serve are calling forth from us, and in us, the very best parts of ourselves.

One of the things that occurs in the hospital all the time is that patients deliver us from ourselves. They pull us out of ourselves. They pull us out of our smaller, pettier selves. They pull us out of our sleepiness and laziness and sluggishness, into a state of alertness and energy and vigor. This is one of the main reasons we do what we do. Because when we are doing it, we are more alive, more awake, more compassionate, and kinder. We are the deeper, better part of ourselves.

But we can't, most of us, get there by ourselves. It's as though we need to be in partnership with someone who is suffering to find the most remarkable selves we can be. This is a very curious dynamic that is not much discussed, but it is an example of the topics we examine in what follows.

The deeper secret undergirding all this talk is that the human vocation itself is, at its core, a healing vocation. In other words, everything that is said about healthcare workers applies to all human beings. It is the spiritual journey of human beings. In the last analysis, the grand themes treated are ones that concern every one of us: suffering, compassion, and healing. In this book, I have chosen to explore these ancient and mighty topics, not primarily through the abstractions and teachings of philosophy and religion, but in the gritty realities of caregiving for those who suffer.

***

Often when I debrief nurses or sit with residents who are working in ICUs, I feel an enormous longing in the room for a solution. Not just a longing, but an expectation. But that is what I cannot offer there, or here: solutions. Indeed, learning to work with moral distress and burnout—with moral anguish and depletion—means getting comfortable in territory where there are no solutions, where there is no fixing.

Lacking solutions, then, what I hope to offer is presence. Presence in which and before which a mutual exploration can take place that allows us to continue to learn and continue to grow, continue to find amazement and energy in the face of great suffering. The presence that offers healing in the face of anguish.

What I promise to do in these pages is to be as present to you as possible. What I am inviting you as the reader to do is to meet me out of the most authentic place in your heart, out of that place that gives rise to your work with those who are in need. I ask you to bring your own presence, your own attentiveness, out of the place from which you come as you step onto the street, into the unit, to the sickbed, as you put your hand on the arm of the person who is suffering, whatever that suffering might be. Together, in dialogue, we can accomplish a great deal.

And, in that spirit, and thinking of all of you who do this caring work, and of your willingness to learn new skills, I want to thank you for what you do every day. I thank you for being willing to step into the field of suffering, and willing to continue to stand there and not abandon. Thank you for stepping forward. Thank you for staying present. Willingness is all. Not everyone is willing. But those of you who are doing this work—especially during this time of the COVID-19 pandemic—have shown that you are willing. To you all I bow, and offer deepest appreciation.

PART I
VOCATION AS PATH

1
The Healing Vocation

Healing is an experience that occurs through encounter. We know that such encounters have occurred throughout human history and across human cultures, across time and across space. I ask that you accept, as a fundamental fact of human life, that the encounter of human beings can, in certain circumstances, result in healing—in a rebalancing, a coming back into wholeness in one's life. We needn't argue about whether healing happens or doesn't happen. We can agree that this is something that human beings experience when they come together in certain kinds of relationships.

I. The Field of Suffering

I begin by making a counterintuitive claim: The field of suffering is the field of healing. What healers do first, what clinicians do, what caregivers do, is enter the field of suffering—and this simple act, all on its own, alters the composition of that field. This simple act brings forth the potential for healing latent there. Let me illustrate this with an example.

Mr. Smith was a homeless man in the hospital awaiting palliative surgery to reduce the size of his brain tumor. He'd already lost his voice and his ability to swallow to this tumor. He communicated by scribbling on a steno pad he always kept with him; he took his meals and, as I learned later, his beer and coffee through his feeding tube. Mr. Smith had been hospitalized once again because his headaches were getting steadily more severe as the tumor expanded. When I walked into his room the first time, I found this wiry, emaciated

Into the Field of Suffering. David Schenck with Scott Neely, Oxford University Press.
 DOI: 10.1093/oso/9780197666739.003.0002

guy sitting shirtless and cross-legged in his bed. I reached out my hand to shake his. He picked this up as a challenge, as a contest of strength. Mr. Smith had a fierce grip, developed over many years of moving about in a wheelchair powered by his arms, and was very proud of it. He also gripped you with his eyes, made more prominent by his weathered, creased face and the wasting from his disease process. Here was an intense man, his suffering written across his body.

When I use the metaphor of "entering the field of suffering," I mean that when we walk into the room of a patient like Mr. Smith, or when such a patient comes into our office or examination room, we enter the field of suffering they carry with them. The suffering that surrounds and suffuses them now surrounds us. In Mr. Smith's case, I was immersed in his homelessness, in the early abuse and horrors of a foster home system that led to his living on the streets, and the substance abuse that kept him there. I was subject to the effects of this tumor that was growing steadily, taking his body from him.

But here you may say, "I don't have his tumor, I'm not homeless. What do you mean that I am immersed in and subject to his suffering?"

Let's shift to a different metaphor for a moment. The patient brings "weather" with himself or herself. They are the center of a storm or a bank of fog, a blizzard or a parching sky. When we enter the room, we step into that weather. We are subject to the climate of their suffering.

Now, go back to the idea that around a person who is ill, there is a field where their suffering is manifest. Likewise, the caregiver who steps into the patient's field brings with them a field of their own. The gift of the person who is the healer is that they bring their field of possibility and hope to engage the field of the person who is suffering.

What does this look like concretely? We go back now to Mr. Smith. At this critical juncture, an unusual constellation of healers entered his life. There was the neurosurgeon and the various medical teams, who were concerned about his decision-making capacity and wanted to give him the best care possible. There was an outreach team serving the homeless, made up mostly of divinity school students, and rooted in a small community church-without-walls that had been following Mr. Smith for many months, visiting him in his downtown haunts. And there was a hospice team that made extraordinary efforts across many months to provide full healthcare for this often very challenging patient. This combination led to Mr. Smith living a very rich eight months after his hospital discharge, despite a prognosis of up to three months maximum survival without the surgery, and three to six months with it. Here the concerted presence of three groups of healers changed his life. These people were able, in effect, to rewrite Mr. Smith's life narrative. His story could no longer be summed up as that of a man repeatedly abandoned throughout life. He was now a man who had been embraced and cared for by friends and healers. In response, he opened to social interactions with dozens of people, where before he had been a resolute loner; he developed a capacity for trust, where before he had dwelt in suspicion. At the very end, Mr. Smith was accompanied through the active dying process by his friends and caregivers, up to the moment of death. He was celebrated by them both before and after that death. This field of suffering had been transformed into a field of healing.

What I propose is that this is a fundamental human process. Some people have always chosen, or been chosen, to be the ones who enter the field of suffering in which others are dwelling and attempt to turn it into a field of healing. This can happen on a very basic level—routine maintenance of chronic conditions or treating sore throats and acute infections. It can also happen, more dramatically, in situations of great suffering and great complexity—the kinds of cases that end up in emergency departments and intensive care units (ICUs). Wherever there is suffering, healing can be offered.

II. The Nature of the Healing Vocation

Given this framework, we can ask in a new way this fundamental question: Who is it that is called to be present in the healing relationship with the person in need of healing? Who is it that steps forward to say, "I will join you. I will be with you. I will stand with you"? Who enters the field of suffering to assist in bringing a broken life back together? We often get bogged down in our thinking: Is healing an art or a science? Is medicine an art or a science? We get bogged down in literature about healing and what makes it happen. What I'm proposing is that we begin with this: "Yes, healing does happen, we know that it does. Let's look at it and see what our experience of it is."

The people who come forward to help, to serve in these situations, are responding to a call. They have taken up something that is a vocation, something that is more than just a job. When I say vocation, what I would like you to think about, or to acknowledge, is that those who have stepped forward to do caring work are engaged in a moral enterprise. They're engaged in an enterprise that enacts a picture of goodness and wholeness, as they join others in trying to make healing happen.

They're also engaged in what I would encourage us to think of as a spiritual venture. Spiritual in the sense of the broadest possible dimensions of meaning and significance in our lives; our connections with our communities, our connection with the universe, with the divine. I also want to use the word "spiritual" to emphasize a dimension of gift, to highlight the fact that there are powers and skills that come forward when we say, "Yes, I will stand in the presence of suffering." There are pieces of ourselves that are drawn forward in that experience that are perhaps beyond what we might usually be able to muster. And, finally, I want to use the word "spiritual" to talk about or to indicate that there's a unique danger here, that we are in the presence of enormous powers and movements in life when we agree to stand with people who are experiencing brokenness.

When I talk about vocation as being more than a job, then, I'm talking about work that embraces all of human being. I'm talking about work that is a deeply felt experience, that engages us mentally. But also, and this is an important thing that we often forget, it engages us physically. Standing in the field of suffering, being in the presence of the sick, of people who are struggling with brokenness, is to put ourselves into charged and challenging territory. This involves a commitment. It involves a commitment to continue to stand with and to act on behalf of people who are suffering. When I say I would like you to think of the work of being a healer, of being in healthcare, of being in a healing profession, as a calling, this is what I have in mind.

I want now to talk about the dangers—I think of them as "spiritual dangers"—that I mentioned just above. One way to think about "call," for those of us who work in healthcare, is to think of our work as responding to the vulnerability of human beings around us. We're stepping forward in response to suffering and vulnerability that we feel as well as see. This, as a component of call, means that there is present most often a sensitivity and an openness to the suffering of others.

This very asset, this very gift, also exposes us to a danger. The label we typically put on that danger is "moral distress." Here I would like to introduce another term, "moral anguish," to complement and deepen the discussion of moral distress. Let's make the distinction between them this way: Moral distress is most commonly understood to be the result of conflict between an individual's moral commitments and the institutional requirements of their workplace. Over time, one becomes increasingly alienated from their moral core. Moral anguish refers to this unresolved conflict within the individual and the crippling pain that comes with it. The ethical self is engaged in an internal battle with the institutional self, as a result of sustained moral distress. This self-division produces moral anguish.

We are constantly in the presence of events, actions, experiences that our patients are undergoing, that our colleagues are undergoing, that are agonizing to us as ethical beings. How in the world can life be like this? How can things like this happen to people? Why do we find such misery all around us? We have the desire to respond to suffering. We have the sensitivity and capacity to do so. And we have, concomitantly, the related danger, which is the potential to be overwhelmed by moral distress and moral anguish. This would be one spiritual danger of the call.

The second spiritual danger is typically labeled "burnout." As with moral distress, I want to introduce an additional term to enrich the discussion. "Depletion" may be thought of as the gradual erosion of one's inner resources and capabilities. It is the sapping of the forces driving the daily work. Burnout would then be the culmination of a long process of depletion, a breakdown of one's ability to function as a professional.

Part of the call is confidence in our ability. If you are a clinician and I come to you, I want you to have confidence in your ability to offer something to me to relieve my suffering, to help me move past my brokenness. I come to you for your surgical skill, your diagnostic skill.

What's the correlation? Early in our careers, I think it is fair to say, ego expansion is part of what we bring to the table as people who step into this vocation. What do I mean by that? It sounds like a purely negative term, but all I mean is that, upon reflection, it is remarkably brazen to step forward and say to someone, "You're in great difficulty; I can help you. Let me offer to help you." We have a sense that we have a capability to alleviate suffering, and this is a remarkable thing. It's a remarkable confidence and an incredibly important piece of what a clinician brings into the exam room.

The spiritual danger of the confidence we have in our ability to relieve suffering is that it can morph or swell into overconfidence, into an inflated sense of what we might be able to do, and this can lead to a state of depletion, a state of deflation. And, unchecked, this can lead to burnout. What happens to us as we move into our careers, move into our development as clinicians, as we see more patients, see more types of suffering, as well as diagnostic conditions that don't respond to the very best that we have to offer? How do we handle that? The spiritual danger here is that as our ego burns off, we find that we don't have other resources to energize us, to move us into creativity in our work.

Given these dangers, I'd like to offer a means to navigate them, a road map—a picture—of what the course of the healer's vocation might be.

III. The Course of the Healing Vocation

Let's suppose that there was a call that, if you reflect on your own experience, you will remember: "Oh, this is when I knew I wanted to be a nurse." "This is when I knew I wanted to be a doctor." "This is when I wanted to go into palliative care." "This is when I wanted to go into oncology." At that moment you had, I'm suggesting, an openness to people's suffering, as well as clear confidence that you could do something about it. At that very moment, you immediately opened yourself to the danger of, or the risk of, the cumulative impact of moral anguish, and of depletion and eventually burnout.

Let's step back and ask, "What is the course of the healer's vocation?" What I want to offer is a sketch, a tentative sketch, of a developmental process or a progression. What we want is a guide that can give us a heads-up, so we are not surprised or thrown by what comes our way in this unusual life. It's a relatively simple framework, with six components (Table 1.1).

I have used the word "phases" to describe this framework to suggest that there is movement, there is progression, as one grows into one's vocation. I have deliberately avoided the term "stages" because

Table 1.1 The Course of the Healing Vocation—Six Phases

1. Call	4. Depletion and burnout
2. Formation	5. Breakthrough
3. Mastery	6. Replenishment and renewal

in my view the progression isn't rigid. One may have intimations of depletion during one's formation period, for example. And we can certainly expect to have repeated cycles of depletion and replenishment.

The first phase is *call* or *vocation*, the willingness to step forward to alleviate the suffering of others—the realization that we can do this, and we choose to do it with our lives.

Phase 2 is the period of *formation*. In this period, we are learning our fundamental skills, and being initiated into the profession. These clinical and diagnostic skills, which I'll talk more about in a moment, are the result of study in medical school and nursing school, and of initial residency experiences. But another set of fundamental skills, which perhaps don't receive enough attention, is the set of skills of relating to patients and developing patient relationships. The period of formation is marked by trials as well. One is tested by mentors, by patients, by peers. One confronts the woundedness of the world on a whole new level—for most people, suffering beyond what they had imagined possible. And this in turn is wounding to the new practitioner. Hence the imagery of initiation that is part and parcel of formation.

Phase 3, straightforwardly, is *mastery*, when one comes into one's own as a clinician. Core technical and relationship skills, and advanced ones, too, have become second nature. One is able consistently to have positive impacts on the lives of patients. This is typically a period marked by feelings of expansiveness and capacity. It can be a period of professional recognition and career advancement, as well as a time of forming deep and satisfying healing relationships with patients and working relationships with colleagues.

Then I want to suggest—and this comes from observation, from personal experience, from listening to practitioners and listening to patients—that there comes a point best described as *depletion*. Here, moral anguish and potential *burnout* loom very large. I want to put this strongly and say that it is virtually inevitable that all of us who respond to this vocation, who choose to stand in the field of suffering, whether that be at the bedside, whether it be in a clinic, whether it be in a community or on the street somewhere, will reach a point of depletion. Depletion is something that is going to happen. We are going to find ourselves facing degrees of suffering that challenge our sense of worth and the worthwhile-ness of all our efforts and activities.

But, going on to Phase 5, I want to posit further that *breakthrough* is possible at just this juncture, and that burnout itself—the breakdown of career or vocation—need not be inevitable. And even if it does happen, this need not be the last word.

There is almost always the potential to move on by being initiated into and mastering a new, different range and type of skills. Skills that will allow you to pick up earlier training and experience and pull it together in a new configuration, and move forward as a clinician with a whole new and different range and type of capacity. Some of these skills may be novel therapies, techniques, or disciplines for working with patients. But the most critical ones, which we will address extensively in this book, are skills involved in working on and within oneself. They could be called meta-skills.

Finally, there is *replenishment and renewal*, that state of having faced deeply one's moral anguish, of having fully addressed one's depletion, of having mastered new professional and personal skills and meta-skills. Vocation is renewed, life infused with new meaning, and the clinician moves forward with novel capacities and a broadened sense of what is possible in the work.

Let me take a minute and talk in more detail about these components. Formation: There's plenty of research on clinical formation and development, which I'm not going to go into here. But I want to touch on research that I did with my colleagues at Vanderbilt Medical Center that highlights healing skills focused specifically on clinician–patient relationships. When we think about a person who becomes a patient, when we think about ourselves when we are patients, we see a person bringing forward their suffering and the disruptions in their life that suffering is causing. But we also see a person bringing forward a willingness to partner with or work with a clinician. And that, most fundamentally, is a willingness to trust. As caregivers, when we step into the healing relationship, we are stepping forward to hold and to carry a vulnerability that our patients typically are not sharing with very many other people in the world. And we're accepting an enormous amount of trust. We're in a remarkably privileged position, and a position of enormous responsibility.

We know from interviewing clinicians recognized by their colleagues as outstanding healers that there are basic skills that tend to be fundamental across a wide variety of disciplines. Our expert clinicians talk first about doing little things, about those first moments of interaction with the patient when you introduce yourself, look them in the eye, if appropriate shake hands, sit down, and then shut up and listen. In our interviews with patients, virtually without exception, patients talked about this as something important to them. Taking time, making sure that you are available to your patients, and that the time that you have with them is time when you are fully present. Time of presence is not the same as clock time. Being in the presence of certain people for five minutes can be far more fulfilling than being with others for an hour. It is the quality of the time that is essential, not the quantity. Our expert clinicians talked about the importance of being open and listening: listening "with the third ear," opening the heart, paying attention to everything that is going on, not just the overt information in a conversation (Churchill and Schenck 2008; Schenck and Churchill 2011).

And there's the skill of finding something to love or to like in our patients, something that engages us. For most all clinicians, it's true, there's a patient who comes immediately to mind who makes them think, "Well, loving 'so-and-so' is going to be tough!" What I would say is that the fallback position is at least trying to find something interesting about the patient, some way to connect. That is to say, "How obnoxious is the T-shirt that Bob is going to have on today likely to be? Can it beat the last T-shirt he had?" The point being to find something, anything, that breaks the person in front of you out of the role of "patient," which in turn can allow you to see them and engage them as a human being.

In short, we have a picture that can be filled out in considerable detail of what healing relationships can look like as they are developed in this initial phase of a clinician's life, in this period of formation.

This research on what makes a healer extraordinary, and what it is that patients have to teach people working in healthcare, was fundamentally a study of what I have called above "mastery." After the research dealing with those studies was completed, I turned my attention to a different element, a different level, of the healing relationship. That is the part where woundedness and the shadow come forward, and we find ourselves struggling with what, if we're going to be honest, can only be called toxicity—both around us and within ourselves. That is, when we agree to take up a life that involves establishing relationships with people who are challenged, broken, in pain, ill, we are embarking on a long and complex series of encounters that can be quite hazardous for the caregiver, as well as for the patient. The patient, obviously, is often at great risk. They have come to you, the clinician, because a critical aspect of their lifeworld is endangered. But you, too, when you take on their suffering, take on their illness. When you step into that field, you are putting your own life—your psyche, your spirit, and indeed your body—at risk.

I began to work with, and mostly listen to, clinicians who were experiencing intensely these specific kinds of risks, and the discouragement and pain they entail. I sat in on and ran debriefing after debriefing in hospitals and other clinical settings with nurses, residents, and both new and experienced attendings. These conversations would inevitably turn to moral distress and burnout, and to frustrations with their institutions and healthcare across the board.

When I talked to such groups about moral distress, I always emphasized that experiencing it did not indicate failure. Moral distress means that you want the right things to be done and done well. The term captures the moral core of the healing vocation, while pointing to the fundamental existential threat introduced by being constantly in the presence of suffering. It is not "mere distress" that can and should be shaken off by responsible grown-ups. It is an agony, a strike at the heart of life and being. And it leads rapidly into the even more painful state of moral anguish. There are certain experiences that are specifically insults to our vocations as healers. All too often, we have the overwhelming sense that our patients and sometimes our coworkers are being treated in ways that compound their suffering and the difficulties in their lives.

Another thing I say to people is, "If you're feeling moral distress and anguish, that's actually great. It means you're human, you're real. It means you have a conscience." In the current context of healthcare, frankly, if you're not feeling moral distress and anguish a significant amount of the time you're on the job, you have either gone completely numb, or you're not paying attention to what is happening around you. We are healers in a broken world.

Given that, it is important to validate experiences of moral anguish as well as those of moral distress. Whereas moral distress can be recognized from the outside when observing a colleague pulled in two directions, moral anguish can be harder to see. It is more personal, a subjective response to prolonged distress, and felt deeper in our being as a threat to core identity. Moral anguish indicates depth commitment to our vocation, and our connection to our patients—the compassion and empathy we feel for them.

We also need to recognize that moral distress and moral anguish move into especially difficult territory when we have the sense of being trapped. When we feel like we are being asked, as bedside nurses often say, "to torture the patient." When we are asked to implement orders that seem likely only to further the patient's suffering, without commensurate therapeutic benefit.

We see things that bring with them a sense of outrage, of violation. It may be something that has happened to a child before they come to the hospital. It may be something that happens to a patient of ours when they are in the hospital. But again, the place where moral anguish turns toxic is most often when it's accompanied by a sense of being trapped or powerless in the face of something that is an offense against our vocation as a healer—or, even more deeply, an offense against our sense of what it fundamentally means to be a human being.

There is very good literature on what happens as experiences of moral distress begin to compound. There is discussion of the "crescendo effect." When you spend time talking with people who work in ICUs, what you find are multiple narratives of morally complex situations and repeated stressful incidents—complex clinical situations, family situations, social situations. The more often one has these experiences, and has passively to endure them, the more often there are unresolved conflicts. And the more often they occur, and with shorter and shorter intervals between them, the less chance we have to return to our baseline, to a position of balance (Epstein and Hamric 2009).

The crescendo effect is one narrative of how depletion, and eventually burnout, begins—this accumulation of distressing events, which is internalized as moral anguish. Let's go back to the two elements of call. We step forward because we want to alleviate the suffering of others, and because we feel we have the ability to do so. In situations of moral distress and moral anguish, we find ourselves unable to relieve the suffering, which is a direct challenge to our call. And, tragically, we sometimes find ourselves doing things that are, in fact, compounding the suffering of our patients. In these cases, both core elements of our call are being insulted.

Again, there is much high-quality literature available about burnout and the symptoms of burnout: loss of energy, loss of creativity, anger, frustration; passive–aggressive behaviors ("I'm over it"); aggressive–aggressive behaviors (surgeons throwing instruments); verbally abusive behaviors. And accompanying those, a numbing of moral sensitivity: "What difference does it make?" An insidious emotional numbing can develop, which can spread—as the life partners of all too many healthcare practitioners will attest—well outside the workplace to become pervasive in a person's life.

What I want to offer is a reframing. What I want to say is, depletion is to be expected in the course of the healer's journey, and career-ending, or even life-ending burnout remains a constant danger. Depletion is an inevitable aspect of the course of the healer's vocation. It will hurt. It does hurt. And one way to think about depletion and burnout is that what we're experiencing as we go through this is the burning off of our ego inflation. It's the burning away, the dissolving, the tearing away, of our sense that we have the capacity to make a difference, that we can change this life, and this life, and this life, and this life. As I said, this sense of our ability to make a change and relieve suffering is a critical piece. It's a decisive part of the call. And yet, it opens us automatically to the painful realization of the enormous range of things that we cannot change.

Depletion and even burnout are also expected because all of us bring to our work, in healthcare and caregiving, unresolved life dynamics that get kicked up or intensified—sore "psychic toes," if you will—that get stomped on in the course of our work. These may be psychological wounds from childhood. They may be traumatic experiences that we've had as adolescents or adults. It may be that we have a chronic illness, a mental illness, or a chronic physical condition. *But it is guaranteed that the pain of those we treat and care for will bring forward the pain that's unresolved in our lives.*

We all have places where we struggle to maintain our balance. We may be extraordinarily skilled and extraordinarily balanced in certain areas of our lives, but in other areas, all of us have struggles. And, in a period of burnout, those struggles can often fill the horizon.

The only real failure in depletion and burnout is trying to feel no pain. Depletion itself is not failure. Burnout itself is not failure. Moral anguish is certainly not failure. Regardless of the institutional talk and motion around this, of administrative efforts made to "relieve burnout" or "bolster moral resilience," we sense condescension and see harsh assessments in conversations with colleagues, supervisors, and the people we supervise. I want to encourage this reframing: Depletion is not failure. Moral anguish is not failure. They are necessary passages for people who have stepped fully and deeply into the vocation of healing. The only failure is trying not to feel pain.

If we can follow the pain into the place of growth and awareness, if we can make the breakthrough, we may find ourselves on the other side, in the state I'm calling replenishment. There we can rediscover ourselves and our vocation. We find that we are more effective than we had imagined possible. We find we have the capacity to stay steady in highly stressful situations that we didn't have earlier in our careers, and the capacity to regain our balance more quickly after these crises have passed. And the way to get to all this is through a process of depletion and breakthrough—the process of the burning away of ego inflation and illusions—and subsequent re-formation and recalibration. Here is involved the development of the more advanced skill sets mentioned above. These skills entail relearning how to do the things that you already know how to do. They are second-order skills—meta-skills, if you will—that experience has shown can help us handle the negativity that we often find ourselves in the midst of. The negativity that is part of being in the field of suffering of others, and the negativity that gets kicked up in ourselves, in our psyches, in our bodies and minds when we do this work.

The core advanced skill is one of heightened, refined awareness. It's the ability to recognize when something negative or toxic is going on around you, and pausing and shifting out of that. What I encourage you to learn to do is to become more and more aware of what is going on in your body, when you are, for example, in the middle of a negative or challenging conversation with a patient, or when you're in the middle of a procedure that is extraordinarily stressful, or when you're in an emergency and feel overwhelmed with the complexity, the urgency and pace, the intensity of what's going on. All of which registers in our body in significant ways.

It's important and very possible to learn how to recognize when our bodies are stressed, and how to change over into a neutral position. Long experience shows that, if we make a basic, simple shift in our posture, it will shift our attention; it will shift our mind, it will shift our heart. It's very straightforward and you can try it right now. Straighten your spine, plant your feet on the floor, tuck your chin a little bit, take a couple of breaths down into your belly. Even 10 or 15 seconds will have an effect. Though of course the more you do this, the more often you practice, the more effective it will be.

There are many explanations for the effectiveness of these movements across spiritual traditions, across medical traditions. We don't have to worry about that here. Just do it. Just try out the practice. Next time you find yourself in a situation that is stressful—a challenging meeting or conference, the next time you're in the clinic—wherever you're standing, wherever you're sitting, feel yourself anchored to the ground. Straighten your spine. Take deep breaths. This is foundational. You're coming back to center.

There are many more skills to learn, and I will discuss a few of them in detail in Chapter 4 and in the dialogues. The critical takeaway is that once we reach a place of burnout, we haven't hit a dead end. We are at a place where we are being forced unpleasantly, usually, to look at things anew. But what I want to promise is, there are new skills to learn, and people to learn them from. There is a new territory of replenishment on the other side of depletion and moral anguish. And we can move into that new territory, if we are faithful to the process, the issues, and the very specific challenges and pains that our burnout puts in front of us. We can move forward into a new phase of our vocation, of our calling.

* * *

I want to close with a single fundamental thought. It's a mantra. It's a principle. And I put it in this position because it is an overarching stance and capability that can be accessed at virtually any point. The mantra is this: "Where there is gratitude, there is healing." Whether it's gratitude for yourself, whether it's gratitude for your patients, whether it's gratitude for the very existence of the world, there rebalancing, refocusing, mustering of new energies, creativity is possible. There healing is happening. When I say it's accessible almost anywhere at any time, what I want to suggest is that you consider that there is always something to be grateful for, even during utterly catastrophic events. If you're in the middle of a code situation, you can be grateful for your team members. If you're in the middle of dealing with a patient who is extraordinarily challenged in whatever way, there is always the possibility of being grateful for the body's capacity to heal or to respond to whatever stress it is finding itself in.

It's possible to practice a variation of what many may know as "loving-kindness meditation," and that is "gratitude meditation." Think of experiences you are grateful for. Think of people you are grateful for. Think of them with regularity. *Build up your capacity for gratitude*. It becomes something that you can tap into at any moment. Gratitude becomes an invaluable resource for renewal.

2

On Depletion and Burnout

Reframing the Darkness

The healing vocation always runs deeper into our lives, and our bodies, than we expect it to, or realize it has. It is a total process—physical and spiritual, psychological and social, as well as professional. It involves more and more experiences of the fierce interconnectedness of people with people, and events with events. And our desire to be immersed in these webs of connections tends to increase the more of these experiences we have. Which ends up meaning that the call to help people in their healing process virtually inevitably becomes a demand that we ourselves heal and grow. The induction into the healing vocation thus has to do not only with our response to the person right in front of us who's in distress but also with our response to our own lives. Vocation in this sense is not primarily about duty or obligation or expectation; it emerges as something essential to the self-realization of a person.

The six phases of the healing vocation presented in Chapter 1 are fundamentally about deepening our lived understanding, our bodily comprehension of what the healing vocation entails as we grow into it, year after year. We are continually circling back around, grasping greater profundity in our initial "yes" to the call than we had the means to realize at the time we stepped into it. As we go through recurrent cycles of depletion and replenishment, the transformation of our work and our lives runs ever deeper—if, that is, we consent to this often excruciatingly painful process. If we do not consent, if we resist and try not to change, try to stay with our old ways and habits, depletion is just around the corner, and

Into the Field of Suffering. David Schenck with Scott Neely, Oxford University Press.
 DOI: 10.1093/oso/9780197666739.003.0003

burnout may not be far away. And once we arrive at burnout, things can take a dramatic turn—as many of us have reason to know.

Picture this: You are the director of an urban free medical clinic providing services to the working poor and the homeless. You spend years serving, fundraising, trying to call attention to the vast inequities in health care access that plague our nation, and trying to help people the best your small staff can. One night, there is a man with a rifle in the parking lot; he's quite angry about not getting the prescription pain meds he wanted. You wrestle him for the gun; fortunately for everyone inside, you win. A few months later, another of your patients—a repeat felon and Vietnam vet—threatens to kill you, after assaulting you at the mental health center. He gets involuntarily committed, but two days later you meet him on the street. The overworked staff at mental health has neglected to let you know he's been released. At this point, a therapist friend says maybe your soul is trying to learn about violence, and makes a friendly, unsolicited referral for you to one of her colleagues. You quit your job to change things up and move to another city, another agency, and keep working to help people. Not many months later, an argument breaks out at your center between a White male in his mid-20s with bipolar 1 disorder, off his medications, having a major manic episode, and a Black male in his early 50s with a crack addiction who, on this day, is relatively sober. The occasion of the dispute is the pair of glasses belonging to the Black man, which had taken six months to procure for him, now broken—inadvertently, maybe, but who can be sure—by the agitated youngster. The older man is known to carry a box blade for cutting, and sharp-pointed barber shears for stabbing; the younger man fills the air with obscenities and racial slurs. You go to intervene. The mental patient assaults you. You black out. When you come to three or four seconds later, you find you have pushed, dragged, shoved this young man, who is half your age and weighs over 200 pounds, the length of a basketball court. You wake up as he is falling backward on his way down to crack his head on the concrete floor. With strength that astonishes, you jerk him straight up by the lapels of his coat and carry him on your forearms the last six to eight feet out the door, where you slam him onto the organization's van. You hold him six inches off the ground as one of your other clients informs you that the youngster is rumored to carry a pistol in his boot. Sirens are wailing everywhere. The police arrive, finding you and the young man equally terrified by your violent outburst. They carry him away. You go to your boss and resign: "Burnt out, I guess," you say to her. She's grateful no one is hurt, is professional, kind, supportive. And accepts the resignation without demur. You spend the next year lugging boxes, loading pallets, and terrorizing the warehouse of the local food bank while learning to drive a forklift.

That's my story, and dramatic as parts of it are, it parallels the stories of many others. After a year of toting boxes and working at recovery, I returned to community agency work, eventually becoming a hospital ethicist, and leading workshops on moral distress and burnout.

We could say burnout has gradations. You can go, I found, from depletion, mild to severe, to burnout, on through to blackout. Participant observation, the anthropologists call it.

I. Depletion Is Inevitable; Burnout Is Not

However dynamic the initial call to the healing vocation is, darkness is obviously and inevitably going to come into a life devoted to attending to the suffering of others. But it is essential to explore the specifics of the dynamics of this inevitability in its peculiar context—the healing vocation and the U.S. health care system. Lacking that clarity, we may be inclined to view periods of darkness, in our own lives and in the lives of others, as failure. But if we can understand that healing is possible only through interconnection and mutuality, that as healers we enter these relationships with patients and clients out of sensitivity and willingness, and that it is this very tightness of the web binding us to those who suffer that is responsible for the periods of darkness, maybe we can hold things differently. Periods of darkness are inevitable if we care, if we are sensitive, if we are willing. Like moral anguish, depletion shows that we are engaged, concerned, and connected.

I will contend throughout this book that recognizing these dynamics, allowing for them, forgiving ourselves for them, is the key to avoiding total burnout and moving past depletion into a new phase of renewal of vocation, and of energy and hope. As an essential piece of that movement, I want to urge a reframing of the freighted term "burnout." I want to encourage using the label "burnout" for a very specific phenomenon: a major hiatus in, or the actual end of, a career, brought about by deep exhaustion of personal and communal resources. This can and does happen, and is terribly unfortunate. It is also not inevitable. Let's acknowledge the power in the imagery of "burning," but try to pick the imagery up in a new way. "Burning" might usefully be thought of as the "burning off" of barriers to further growth as a caregiver. Perhaps the ego inflation that often comes early in our careers. Perhaps things extraneous to the vocation—like money or prestige—that seemed important when we first started out, can be set aside, because they are no longer nourishing. The best thing about the word "burnout" is that it generates images of something dire, something that we need to remedy, and quickly!

And there's the problem of the term's severely judgmental component. When people talk about someone being "burned out," they're usually giving us the picture of a person whose behavior reflects frustration, impatience, a certain harshness, a lack of empathy or compassion, disconnection. There is almost always a sense that multiple failures are involved, whether this is spoken or not. None of this baggage is helpful in the context of health care, or indeed anywhere else in life.

As a critical next element of the reframing, I would urge attention be given to the "depletion" that commonly precedes and often results in "burnout" proper. Think of this vocational depletion as analogous to what in spiritual literature has been spoken of as "the dark night of the soul." This a period where the resources that you had in the beginning of your career—or at the beginning of your spiritual life, in the case of the dark night of the soul—that nourished you for a long time and kept you going, and gave you energy and direction, have dissipated. You're barely hanging on, mostly by virtue of willpower or determination (John of the Cross 1987).

One way to describe "depletion" would be to say that our core energy steadily declines. And we need to recognize that there's a specific terror that goes with that. The experience is that our core energy is no longer adequate for us to handle what is right in front of us. We are speaking of that very core energy that has gotten you through all your formation period in your profession—your training and your residencies, say—and sustained you in your time of mastery. This is a crisis of confidence in that core energy. That flame itself seems to be flickering or going out.

Professional demands multiply, from patients, from colleagues, from administrators and auditors. We find ourselves challenged in new ways, and our old motivations aren't up to the new realities. Sometimes our old therapeutic approaches don't work. The assurance that we can resolve crises slips away. Levels of discouragement rise exponentially. Moral anguish intensifies. There is plenty of excellent academic, professional, and self-help literature on this. Which means: This a common experience. We know what this looks like; we see it all around us, we feel it in our own lives (Epstein et al. 2019; Hamric, Borchers, and Epstein 2012; Katz and Johnson 2006; Rotenstein et al. 2018; Rushton 2018; Ulrich and Grady 2018).

All that is difficult. But if at the core, the energies that have kept us going this whole time are faltering—what then? If, suddenly, they begin to flicker and we think they're going to go cold, we are truly in trouble. This is where we have people leaving the profession. We have abuse of patients and colleagues, we have addictions, we have suicides.

What I am proposing is that there can be a different outcome if we can mark the evidence of depletion not as a failure or a dead end, but as a clear and decisive sign. A sign that it's time to recalibrate, time to reassess what resources one has available, and what purpose one finds in the work. Time to recognize that one of the main reasons that the old strategies and old motivations don't work anymore is that the practitioner, the caregiver is *outgrowing* them. That doesn't mean failure. It means that potentially, one is in a place to learn a whole new level of skill and find a whole new vein of richness in the work.

For instance: A skilled clinician had worked as a licensed therapist for 10 years with college students. She was very good, well respected. Then her mother died a terrible death, following a two-year illness with cancer. After that, it became incredibly burdensome to keep talking with students about their depression and grief, while she was immersed in her own depression and grief. After extensive discussion with her partner and her friends, she decided to take extended leave. For two years, she did not work. In the first year, she did extensive research on nutrition and food as medicine, body–mind–spirit wholeness, and wellness. She found a certification program in health coaching. In the second year, she did that program.

Emerging from that time of profound redress and reevaluation, she is a therapist still. But alongside the counseling she had always done, she has added her new training in nutrition and wellness. She has a much more eclectic and integrated practice. There is an integration of what had always been a profound vocation and skill set with new skills and ways of doing the practice. It's like a new person emerged.

Other cases come to mind as well, drawn from interviews with extraordinary clinicians:

- A physician who left private practice in family medicine to direct a residency program in family medicine, taking the utterly outstanding clinical skills he had developed as a practitioner, and putting them to work training residents—a multiplication of his gifts that became very satisfying.
- A nurse practitioner who began to feel that what she was called on to do day in and day out missed certain fundamental components of what constituted real health. She went back to school to become an acupuncturist and was able to develop a thriving practice, combining insights and therapies from Western allopathic medicine and Chinese medicine.
- A craniosacral practitioner who was, in her own words, a "crank-and-yank physical therapist," who went through a remarkable spiritual experience and afterward was able to sense people's pain, and the challenges going on in their bodies, in new and remarkable ways. She went back to school and got additional training in craniosacral therapies.
- An intensive care unit (ICU) nurse who got intrigued with and deeply concerned by the rescue ideology dominant in U.S. health care, went back to school to get training in anthropology, wrote an outstanding book in the field, and became a practicing ethicist.

And then there were those who did not change fields or careers, but found ways to renew themselves within their current work settings:

- A nurse working in a pediatric ICU who got additional training in the use of extracorporeal membrane oxygenation (ECMO) machines, allowing her to care for her patients on a whole different level, and giving her access to other units in the hospital where she could work with adult patients and specialty teams new to her.
- A psychologist specializing in treating victims of trauma who pursued additional training in therapeutic approaches to

dissociative identity disorder, in part to satisfy intellectual curiosity and keep her mind growing, and in part to understand trauma more thoroughly, and expand the range and type of care she could give her clients.

- A nurse working in oncology who, as her understanding of the dying process deepened, shifted her practice to an ICU setting, then to a palliative care team, and was eventually trained as a death doula.
- A neurologist who made a return to one of her undergraduate majors, philosophy, when she took up the position of medical director for the ethics consultation team in her hospital and began reviewing cases from both medical and bioethical points of view.

It is worth noting that none of the practitioners in these two sets of cases had progressed to burnout. It is also worth noting that, for each of the practitioners in both groups, a willingness to explore something new, to push beyond existing habits and routines was required for them to make their breakthroughs. In each case, the practitioner became aware of mounting dissatisfaction, distress, and depletion and made changes aimed at replenishment and renewal before full burnout occurred. The essential realization is how much is possible, once we have been chastened and opened by the experience of darkness and barrenness, by the dive into depletion.

Let's turn to examine specific features of health care and healing that make such depletion virtually inevitable.

II. Institutional Vectors

When I use the term "institutional vectors," I mean to indicate economic, logistical, and administrative features of our health care system that have an impact on depletion and moral anguish. The COVID-19 pandemic has made abundantly clear to the world at large what people in hospitals have long known: Our health care system is virtually designed to produce depletion and burnout. The emphasis on profit over people—profit over both patients and staff—is the decisive vector. The overwhelming workload, the inadequacy of safety protection for staff and for patients. The lack of control over one's working conditions. There is the fact that the lowest paid staff members, who tend to be women and tend to be people of color—nurses' aides and maintenance workers—are subjected to the most dangerous conditions with the least protection. I will return to each of these points.

As of this writing, the pandemic has been with us more than two years. It is useful to go back in time and remind ourselves of the progression of mounting challenges for health care staff raised by the rampage of COVID-19:

Safety: Do we have enough personal protective equipment—including very basic things like masks and gowns? Do we have enough testing capacity? Do we have adequate protocols? Can we care for COVID patients without getting sick and without dying?

Shortages: Almost immediately supply lines were overwhelmed. There were not enough masks, gloves, or disposable gowns. Not enough tubes for ventilators. Not enough ventilators themselves, early on the most crucial machine for treating COVID patients. In some regions, there were sporadic shortages of dialysis machines and cardiac monitors.

Overload: As the numbers of very sick patients accelerated rapidly, ICUs became full and overflowed. The sheer volume and acuity of cases were exhausting staff. And staff members in great numbers were getting ill themselves, which produced even more staff shortages, even more overload. Morgues filled. Mass graves were dug in New York City.

Layoffs: Counterintuitively, as units and teams were overwhelmed, many hospitals laid off enormous numbers of staff. These were primarily in two categories: (1) Medical staff who were performing or supporting elective procedures, not needed because such procedures were mostly prohibited by COVID protocols. Often these staff members did not have the necessary skills to be reassigned to care for COVID patients or to work in other ICUs. (2) Support staff, chaplains and social workers among them. These could certainly have helped with COVID patients and their families. But without a significant volume of elective procedures, there wasn't enough revenue in most hospitals to keep a full complement of them on the payroll.

Emotional labor: Because families were not allowed on COVID units, hospital staff, and especially nurses, were called on to do enormous amounts of emotional labor, above and beyond what they have traditionally done. No one else was there to comfort these patients; no one else was there to talk with them about dying or listen to their fears. And there was the sheer magnitude of grief: The numbers of deaths experienced in short periods of time ran well beyond nearly everyone's experience.

Backlash against public health measures: Masks, social distancing, and eventually vaccinations were and are resisted by significant portions of the country, varying by region. This burdened hospitals with new surges of very ill patients during the very months many professionals had expected to be able to devote time and energy to healing themselves, and had expected the systems supporting them to be recovering. Instead, providers have recurringly faced record numbers of cases. The frustration and sense of futility among health care workers around these dynamics can hardly be overestimated.

Staffing difficulties: Too many of the best died; many got very sick; many remain ill. Many simply quit. And many transfered to new locations where the pay is far better. All this contributes to shortages and team instability in a time when demand for services and skills is skyrocketing.

These points were drawn from the headlines of national and local newspapers, from anecdotal accounts of frontline workers, and from studies of the history of the pandemic beginning to emerge. For all it has obscured, the COVID-19 pandemic has shone a bright light on fundamental structural flaws in the U.S. health care system.

But none of these "revelations" was a surprise to those who have been studying depletion and burnout in its institutional setting. The literature in this area is burgeoning exponentially, and I will not attempt to survey or summarize it. A useful overview may be found in Chapter 2 of Cynda Rushton's landmark study, *Moral Resilience* (2018; see also Epstein and Hamric 2009; Jameton 2017; Rodney 2017; Vercio et al. 2021). For my purposes, focused as I am on frontline workers, and the moral and spiritual dimensions of the vocation of healing, it will suffice to name five major institutional vectors that contribute directly and powerfully to moral anguish, depletion, and burnout. It is no exaggeration to say that these five vectors together produce a system that consumes the best energies of our best providers and induces burnout in them. In the sections that follow this one, I will consider more subtle dynamics of the culture of health care and the nature of encounters with the suffering, which also contribute to depletion and burnout. Here I am dealing with the obvious ones, the ones everyone talks about:

Profit over patients and staff: Economizing, cutting costs, eye on the margin or the bottom line. Constant understaffing. Inadequate supply lines, shortages of safety equipment, and neglect of safety procedures. The pressure to perform less-than-necessary, but highly profitable, procedures. (As one colleague very memorably put it: "Prioritizing dollars-per-second over care-per-patient").

Powerlessness: Who makes decisions, the everyday ones, and the systemic ones? Power hierarchies in hospitals. Pervasive sexism and racism. The changing roles of patient families. Very limited authority to make changes in procedures, protocols, and policies that directly impact patient care and staff safety.

Serving technology: Nurses' days are commonly driven by the needs of machines linked to patients, as much as or more than by the needs of the patient. Heavy burden of charting for physicians and nurses. The electronic medical record is the cash register; it is the legal record of care given; it is also one place where continuity of care and coordination among teams must be accomplished.

Futility: Hospitals sit in the middle of an unhealthy culture. Safety net hospitals have become virtually the only resource for many of the most vulnerable in our communities. Here we find our mentally ill, our homeless, and our domestic violence victims, along with the acutely ill, the chronically ill, and the episodically ill. The "revolving door" cycle of discharge and readmission for many patients who are among the most challenging medically and/or psychosocially. The sense of being overwhelmed by nonmedical needs, as well as medical ones.

Job satisfaction: Repetitious tasks, as percentage of workday. Low pay. Lack of respect by the public, patients, and administrators. Lack of institutional resources. Sheer fatigue; crippling exhaustion; moral distress.

There are also depleting institutional dynamics at work in health care settings outside the hospital. Practitioners throughout these systems report

- too many patients, too little time;

- innumerable patients who would benefit from interventions and procedures not covered by Medicaid, or severely limited by other types of coverage;
- racial and economic disparities in health care outcomes, in the types and quality of care received;
- patients who cannot afford or do not have access to basic medical care, who in the short run may do well enough, but in the long run will experience fatal consequences from conditions such as hypertension and diabetes; and
- repetitive, monotonous data entry and clerical tasks that have little to do with actual patient care, and everything to do with meeting regulatory and insurance requirements.

These institutional vectors also intersect in ways that compound and amplify their effects. All together they rank very high as causes of burnout, depletion, and moral distress. For this reason, there is a dialogue in Part II dedicated to staff advocacy in the workplace. There I discuss why learning to have an impact on one's institutional environment is essential for coping with depletion and moral anguish. But as the following sections in this chapter will show, there are other dynamics involved in caring for those who suffer that are major contributors to burnout as well.

We must acknowledge, though, that institutional change, as necessary as it is, is going to take quite a long time and will involve arduous labor. We can be overwhelmed by hopelessness, by the scope of these vectors, and the complexity and inertia of these systems.

How then can we maintain ourselves while that change is happening? Furthermore, who are the change makers? Do we want those to be the administrators concerned primarily about lost days of work and balance sheets? Or do we want those leaders to be people who are committed to their vocation, and who are exploring and growing in their own lives?

I argue that the spiritual and personal exploration urged in this book is essential for sustaining quality institutional change.

Otherwise, that change is likely to be dominated by administrative priorities. Or by the sheer reactivity of the most powerless. Both of which can, as we know from political, economic, and more specifically labor history, often lead to changes in systems that further harm the activists and the powerless. We can and must do better than this. The healing vocation, our patients, and our colleagues demand that of us.

I turn next to more subtle aspects of the vocation itself that will help us understand better how to make lasting change in our institutions, and in our own lives.

III. The Limitations of Mastery

The exhilaration of the mastery of successful therapies and interventions gives us core confidence: With the right group of people, we can handle almost anything that gets thrown at us. We have encounters with patients and families, where remarkable things happen, and we come to believe in our ability to make change. Change, which is astonishing, given the dire nature of many of the situations we are called into. The work can be transformative.

The heroism of healing and rescue is a powerful thing (Chapple 2010, 2015). It's certainly powerful for the hero, for the rescuer. And it's a powerful experience for the person who has been rescued, the person whose life and health have been saved. There's the sheer amount of gratitude that one has as a patient—having been besieged in body and mind, having someone show up to assist at that critical moment, and coming out on the other side of a crisis.

We in health care can get an overinflated or overly grand sense of ourselves and what we're able to control. This is a regular theme in mythology; it's a regular theme in literature; it's a regular theme in spiritual literature. We get intoxicated with what we're able to accomplish. We tend to forget that everything that happens is multifactorial, is the result of many causes. We have, instead, a scientific method guiding our work that's based on an effort to identify relatively small sets of causes or indeed a single cause. Most of our literature in medicine, in social science, and even in ethics is driven by the idea that we can identify a small set of causes, and we can have strategies that replicate success. But the truth of the matter is that the web of causality is so extraordinarily complex, that the idea that we might be able regularly to identify or control causes is self-deceptive at best.

Despite that, what we continue to do is say, "Oh right. This is the single cause, or these are the two causes, so we'll fix these, and everything is going to be okay." If the tide is right and we're in the right season and things conspire that are way beyond our grasp, then that does happen sometimes and it's wonderful. Other times it's not going to happen.

One of the things that leads to depletion, to the arid place, the desert place, to the place where it feels that nourishment and life are being sucked out, is this core human level of fear and puzzlement about the intractable nature of human suffering: "Why is the world like this?" "Why is it that children are tortured by disease?" "Why is it that there are certain families that have certain patterns and yet nobody seems to be able to change?" "Why do certain communities suffer a vastly disproportionate amount of injustice?"

We develop technologies; we develop strategies. We develop policies; we develop systems. We develop therapies. We develop all kinds of solutions and interventions intended to relieve suffering or alleviate the worst of suffering. But there comes a time in the development of a healer's career when the irremediable nature of suffering, the inevitability, sits there staring you in the face and it's clear that it's not going anywhere. We become aware that the complexity of causality amounts to the inevitability of failure, even in the face of confidence and skills and intent to bring about results that are healing.

There are many spiritual and religious traditions and teachings about the centrality of suffering in human life. For most of us who go to work in healing vocations, our intention and our hope is to relieve suffering. Our hope is that suffering can be avoided or set aside. What we come to learn is that this may be true for a portion of the suffering we witness, but for another portion it's just there and isn't going away. How do we learn to live with the fact that suffering is just there? How do we talk to our patients and teach them that there are certain components of suffering that are not going to change, that are going to be a permanent part of their lives?

IV. What to Do with the Unfixable?

One of the interesting and incredibly difficult things that's happened in health care is that we have become too good at producing miraculous results. When you say to a family of a patient who's in the ICU, "We can't do X, we can't do Y," most of the time you're met with disbelief. "There must be something you can do," people respond (Kaufman 2015). Or people are on the internet searching, looking for answers, with the sense that there must always be an action or a cure—an action that can alleviate or a cure that's available. It's very exhilarating to be the person who delivers that cure. It's exhilarating in its own way to be the person who is expected to fix things. But the process of trying to teach a patient, a family, a client after there's nothing else that can be done involves a dismantling of the expectation that is there, just there, in the health care world.

And this is an expectation that has been placed on you. Sometimes of course this draws the anger, and the rage, and the frustration, of a family, of a patient, of a community that was hoping for a solution and in fact that felt like you were going to be the person who delivered that solution. And that initial expectation turns. The medical team, the doctor, the therapist, the person who's going to deliver the solution has failed. This is a confrontation with the inevitability of suffering, for you and for the people with whom you are working, the people you promised to accompany on their journey. You have participated in a process of hopefulness that then, as it evaporates, can be very challenging and very painful. It's not like these things aren't happening before the burnout period begins. But we're younger then. We're more resilient. We haven't seen as much. "The body is keeping the score"—and as you get older, the body is *still* keeping the score, still recording the emotional blows (Van der Kolk 2015). If we don't learn to move the negativity, and move with it, eventually the accumulation takes an enormous toll. We find ourselves, all of us, feeling constricted, feeling afraid, feeling resentful in the very territories where we had felt inspired before.

These are profound and terrible moments, which raise questions our training has typically not prepared us to address: What is the nature of suffering? Of causality? Of our connection to others? What is the nature of service? Of sacrifice? Of compassion?

So much of our journey toward mastery—graduate school, medical school, internships, practicums—is all about challenging us to push our abilities further, to expand our understanding, to implement that expanded understanding, to defy fatigue. We are trained to exceed our limits. We have been schooled constantly to push past whatever could hold us back.

Given all of that, consider what it is like for us to stand with a patient or a family and say, "There's nothing more we can do."

First, there's a human consequence, and perhaps a tragedy that will continue unfolding and that we cannot prevent. That's the first thing, which is enormous.

And there's a relational challenge that must be addressed. How am I going to communicate with these folks?

Finally, it seems that our training has left us ill-equipped to deal with the very fact of limitation. The practitioner shoulders in such moments the limitations of all of medical science: "I'm very sorry, but medicine has gone no further than this at this point. Your loved one has moved beyond that boundary." And there's not much training or wisdom imparted about what we should do when the people we're working with have gone beyond medicine.

This brings to mind a singular family meeting in a neonatology unit with a couple I had spent several weeks working with and liked very much. They had come to trust me and my contributions as they had struggled to shape the goals and course of care for their son. We were working well together. But finally, we got to a point where there was nothing left to do, in terms of further medical interventions. At this juncture, in the judgment of the team, the best thing to do was to withdraw the life-sustaining therapies (for context, see Schenck 2015).

I remember clearly the final meeting with these parents. I said to them, "We feel like what we're doing for your son is harming him. There's no way forward. It's time for us to stop." The father was sitting right next to me. He looked me straight in the eye and said, "Are you saying there is *nothing else* you can do for my son?"

My answer was—it had to be: "Yes, that's what I'm saying. There is nothing we can do."

He got up suddenly and walked out. Totally understandably. But then, to his credit, he called me back an hour and a half later and said, "I couldn't sit there. I didn't mean any disrespect. I appreciate what you've done, and I appreciate what the medical team had done, I just couldn't stand it." In truth, it was quite remarkable—and deeply generous—that he made the call.

That unbearable moment is like sitting in a small boat at the edge of the world and looking at the water pouring over the side. The father is saying, "Let me row back a few strokes." He's saying, "Can't we back up a little bit?" And what you are compelled to say is, "No." But you can add: "I will stay right with you. I'll stay right here and look over the falls with you."

There are many clinicians who are not willing to do that. Practitioners don't like to have those conversations with patients. Many get very skilled at avoiding them. And because it's such uncomfortable territory, clinicians end up hiding behind their technical language, and in consequence these conversations often go very badly. We hide behind the competencies we do have to disguise what we cannot control. To have the bald facts of our human limitations out in the open is something that many practitioners are afraid of.

But it's important to note that there can also be a discovery or an expansion of resources when we stay in that madly uncomfortable place—that place where medications don't seem to be working, where a medical apparatus doesn't seem to offer explanation or an open door, where no therapeutic procedures are feasible, where clinical trials have failed. Where all the fancy techniques aren't working, the simple move of compassion still can work, and indeed in this moment compassion has vital work to do. When you look at a person and say, "There's nothing more we can do," you're not saying, "Therefore, I opt out of this relationship." And by doing the opposite, by holding the paradox—"There's nothing more we can do, and I will be there"—you access a whole new range of relational, presence-oriented, compassionate resources that may not have been apparent when technical options were more readily available.

It is the case, after all, that the bulk of Western, allopathic medicine is about the refusal of the ultimate; it's about sustaining, maintaining, or at worst coping. And that creates a system of thought that in turn creates systems of medicine in which we are called to work. But all of that is in a certain sense choosing temporary solutions to hold back the inevitable.

That place where there's nothing to be done is the place everyone is afraid of. And it is a terrifying place. It's terrifying, because even if you're practiced in it, you have no idea what's going to happen. It is a place that is absolutely relational, and in that relational space you're giving the worst possible news to a person. Emotionally, even if not objectively, it is the worst possible news. The sudden expansion of their world is in that moment a terribly painful experience. Yet, if it's going to happen in a way that can yield growth in the future, people need to be present, and they need to stay present. It's a question of how much people can hold, and how much we can hold, as we encourage people to stay put in the pain. But if we are afraid to go there, and can't stay there ourselves, we have no right to ask this of someone else.

V. Confronting Institutional Narratives

Understanding the depth of our interconnections with others, and especially our patients, is such a challenge in health care in significant part because there are so many competing narratives for interpreting the bodies in front of us. And very few of those narratives have to do with the personhood of the patient. We have lab metrics of all different sorts, blood gases, blood cultures, biopsies; we have multiple sorts of imaging of this, that, and the other. We have diagnostic categories and trajectories and prognoses, and disagreements about diagnoses and prognoses. And they all provide compelling and authoritative narratives, backed by science and machines and data, those ostensibly objective markers.

One reason these technical narratives, as well as the plethora of negative and judgmental narratives about patients, end up being compelling is that they're a way for us to deal with our own fear and anxiety. They're a way to block out the very pain generated in our bodies by witnessing or hearing what's before us. Yet, paradoxically, these distancing narratives contribute to vocational depletion by disconnecting us from the personhood of the patient. They obstruct the relationships with others that are essential nourishment for clinicians and caregivers.

And there's the fact that we're often operating in a setting where things happen fast, where there are many patients, where many bodies are present. How can we find a way to continue to see the person with whom we are working? Let's consider two very specific tactics, as a way into this:

1. At the bedside of a person, try stopping for a minute and placing your hand on their arm—and have your breathing follow their breathing, whether they're apneic, whether they're hyperventilating, whatever the breath movement is. Attempt to synchronize the movement of your chest with the movement of their chest. You can do this in 30 seconds. If you do it for 20 minutes, that's a very different experience—an extraordinary experience! But it doesn't take long for a direct, somatic encounter: "Oh yes, this right in front of me is a human being who is breathing and struggling to breathe and struggling to stay alive, just as I am." Connecting our breathing to theirs offers us a window into a core element of the patient's existence and a core element of their humanity.
2. We can ask families to tell us about the patient. Consider an ICU setting with a patient who can't answer questions. Sometimes people will say, "Do you want to know their medical history?" And it's good to say, "No, I've heard their medical history, and I do want to hear your version of it in a minute. But first tell me about your daughter. What is she like? What name does she go by? What is important to her? Would you mind showing me a video or picture of her?" "Tell me about your father. What was he like, before coming here? What did he love to do? What was his favorite thing? What was the best thing that you did together?"

What comes next depends on who the person is and what the conversation is like. If they say, "He was a sorry-ass fool!" we can pivot to keep opening our connection. We can ask: "What was the worst thing about him?" The point is to allow the family member to give you a narrative that is different from the ones being pushed onto the patient, in the context of a hospital, in the context of the ICU.

A critical element of seeing is the ability to bracket the narratives that are being pushed on us in a healthcare setting. Mostly these are medical and disease narratives. But sometimes there are moral or even religious narratives, often very judgmental narratives, that are placed on patients and on families by staff. Narratives like "This family never shows up," and comments or speculation about why this happens. Or: "Why does this person have this condition," followed by a story about how it is their fault. Judgmental narratives like these are, unfortunately, constantly in use.

It is essential to develop an ability to dismantle habitual and reactive narratives and receive new information. This will mean becoming aware of the pressures to apply certain kinds of narratives in specific work settings. If we work in a community agency, we should look for alternative narratives to apply to specific neighborhoods or streets. In clinical settings there are multiple narratives paired with certain families, and varying diagnoses for mental and physical illnesses. How do we resist that "obvious" narrative and grasp the story of the person who's in front of us as it is told by them or by their body?

This pushes into deeper territory, into somatic territory, where our fear and our anxiety, as physical reactions, can also prevent us from seeing. So, another important skill for continuing to be able to see our patients, for continuing to see the brokenness around us, is the ability to deal with our own anxiety and our own fear, and the physiology of it. Not just beliefs about it, but action. Are we able to calm ourselves? Are we able to use breathing techniques? Are we able to use centering and mindfulness practices to keep our body in a calm enough state that we can receive new information from what is going on in front of us?

As I will explore in Chapter 3, this comes down in large part to being aware of our woundedness, and aware of the reasons for those quite specific elements of our work that give us difficulty. For instance, if we are in settings that involve a potential for violence and we feel threatened by what's going on, and we have anxiety through the ceiling, we need to be asking ourselves what our past experience of violence is. Is there personal history there? We don't necessarily have to have grown up in what would be considered a "violent" neighborhood or region to have had considerable experience with violence. Many households that appear quite tame are quite violent. Our whole society, and the media culture reflective of our society, is enormously violent. We all live immersed in it. Or perhaps there are situations of economic distress that generate enormous anxiety that make it hard for us to see what's going on?

The question about seeing becomes extraordinarily complex in terms of making a space for the narrative and the life and the reality of the person in front of you to come through the institutional and cultural narratives surrounding them. But there's also your honed internal skill at clearing up the narratives that are in your own mind and that are written in your body, so you can be in a position to receive the person given over into your care.

* * *

The work that the healing vocation results in is life giving; it is work that can be energizing and deeply animating for all involved. Enormous human power and energy are often tapped and released as healing occurs. Yet it's critical to emphasize that this release occurs in encounters between individuals, and in the relationships that develop from these encounters. The call isn't just about you, the clinician, the practitioner, the caregiver. It's about what happens when your passions and your gifts meet a person and a situation that elicits them, brings them forth, pulls them out of you.

What's always challenging, though, and often infuriating, is that the calling addresses us on many different levels at once, and from so many different directions. This can, at times, be a great gift to us. We find traits, abilities, and capabilities in ourselves that we didn't know that we had. We find sources of animation and enlightenment that we weren't aware of in our lives, our families, our communities. We find an exhilaration and an intoxication in meeting others in places of great distress, in being compelled to speak honestly and act directly, in solving complex problems under terrible constraints. To touch the person whose body is distressed, touch the person who is dying and isolated, to reach that person no one wants to touch—this involves full recognition of the depth of the challenge of what it means to be alive, of what it means to be human.

Yet we know we're not going to end illness, end disease, end suffering. We know, in truth, that all these are, for all practical purposes, infinite in extent. Say you work in an ICU, and you have one person who's very sick and either they get better, or they die—and then you have another person who's very sick—and then you have another person who's very sick—which means you are, in a certain sense, experiencing permanent failure.

True, we must get beyond the felt assumption that death equals failure. We must come to know more and more deeply that failure is, rather, lack of dignity and comfort and care. But one question embedded in the call to the healing vocation is this: "Are you willing to come back to this again and again and again? Knowing that it is endless, knowing that it is in a certain way infinite, does that make it futile?" If we're honest, we'll admit that these underlying questions are there in the very beginning, when patient and caregiver join forces and say, "There's this impossible thing going on, and we're going to take it on together." These are questions that return, and that will manifest themselves poignantly in moral anguish, in depletion, and, if left unaddressed, in burnout.

3

How Breakthrough Happens

The Mutuality of Healing

We have been exploring together the call that people feel to work with others who are in situations of extreme suffering and brokenness. In this chapter, we want to question ourselves more intensely as to the "why" and the "how" of our work. A portion of it, to be sure, is generosity and sensitivity and attentiveness to the other. But if we are honest with ourselves, we can also become aware of the role our own life challenges, our own vulnerability has in our work.

The person for whom you are caring at any given moment is carrying a certain portion of human suffering for all of us. This makes for a claim on our generosity and should inspire gratitude. But, in addition, they are there as a projection screen for our own challenges and difficulties. They are there like a tuning fork, which echoes or vibrates in sync with us—their suffering with our suffering. Diagnosing the one in front of us is a way for us to engage indirectly in further exploration of our own suffering. Our deepest and most difficult wounds can often only be approached through another's wounds. In other words, there most often is a mutuality, a reciprocity, in healing. As we find ways to offer healing for our patients, we are often also beginning to heal ourselves.

The very fact that I'm engaged with this particular patient may be a recognition that they offer gifts, traits, or skills that I need to learn from. Part of the ongoing work is to recognize, in each situation, what the gift or lesson the person in front of us is offering. They could even be lying comatose, in a persistent vegetative state, and still be teaching us. Maybe we suddenly understand arthritis pain

Into the Field of Suffering. David Schenck with Scott Neely, Oxford University Press.
 DOI: 10.1093/oso/9780197666739.003.0004

in a new way or recognize humanity in an addict in a way we hadn't before. Which in turn allows us to see the person living with addiction in our family in a different light.

Each patient encounter thus has the potential to be a mutually healing process. There is also, to be sure, the darker side of this, which is that a mutually wounding process can occur. But this is one more reason to begin to work with our own woundedness. We don't want to end up in an unconscious dance that is further damaging for our patients and ourselves.

That there are gifts constantly being manifested in every person who comes before us: try taking that as an assumption, as a method, in every encounter. This can become a critical resource and a strategy for dealing with being in challenging, potentially depleting situations. And the opposite assumption, if you will—that I'm the person who's always giving and pouring myself out—that, in and of itself, is a guaranteed recipe for burnout. It's a narrative that is going to drive us quickly into a place of debilitating exhaustion. And, although totally understandable—and something we all feel at times—it's also a fundamental misreading of the basic encounter. If we can see that the other person—the patient, the client—offers us insight into our own lives and an awakening of our deeper self, and at the same time that we can offer something that is of benefit to them, then that's a very different narrative.

I. Wound as Teacher, Wound as Resource

Suppose you're a person who chose a helping profession, a serving profession, years ago, but you're worn out by it. You pick up this book, and as you read, memories stir of how things were when you started out; you recall when you were certain you could make a difference and be a benefit to others. There was exhilaration; you did your work well. It was fulfilling; it was meaningful. Yet, whatever the initial motivation and energy was, you're worn out now.

And now I'm telling you that you got into this work because of some kind of wound. It would be reasonable to object: "What are you talking about, wound? Why do you say I have a wound? I'm the one who helps. Why is this important?" And I would say, "Great questions. This is where the deep work begins."

The first thing is to linger a moment over the imagery of "wound"—a deliberately unsettling word choice. There is the vividness of what is called to mind, as opposed to the more abstract term "suffering" or the mild term "distress." We are in the territory of injury and trauma, but deliberately avoiding those more technical terms. Just as important, there is the mythic layer associated with the pairing of wound and healing. In Western culture, the stories of Jacob and of Tiresias, of Parsifal and the Fisher King, and the many variants of Christology, form the background. Accounts of shamanic initiation in Indigenous cultures emphasize the woundedness that marks out the candidate for their vocation. Very often, one who bears a wound is called to heal the wounds of others: the wounded healer (Frank 2013; R. Johnson 1995; Nouwen 1979; Eliade 2004).

The next thing is to recognize that if we think that all wounding in any encounter, in any relationship, is all on the other side, we are in deep trouble. As is our patient!

True, the person who's sitting in the waiting room waiting for the doctor to come in is not thinking that the doctor or nurse might be vulnerable, too. They're thinking the practitioner is going to be able to offer something to them that's going to help them out. And that is appropriate. But the underlying connection—human connection, human recognition, human trust—that sense that we're in this together—is totally essential for any healing work. And it ideally involves awareness on both sides that there is shared vulnerability, and shared capacity and capabilities.

We can hope that the physician or the nurse remembers that the patient has capacity and not just woundedness; and we hope that the patient and the family remember that the physician and the nurse have wounds, too, and not just capacity. But we often go off course and end up getting captured in a place where we assume that some people have all the capacity, and others have all the woundedness.

It's worthwhile to look carefully at the nature of the connection that is present when we say we are feeling "compassion for another person." What is that connection? In large part, it is the shared sense of the challenge of being human. The shared sense that we would all prefer not to be suffering. Grounding ourselves in that, recognizing that we all have suffering, of which we'd prefer to be relieved—and that certainly the other person has that as well.

Simply being aware of what one's own challenges are can help in making connections with other people. My experience is that when I'm dealing with someone who is in great distress, often a key element in establishing trust is the fact that they recognize that I have been in a place of great difficulty myself. And this doesn't necessarily come from sharing that overtly. There's a moment in the conversation in which it registers that we're in this together. It's the recognition that I'm not there as the person who has the solutions or the answers, which I'm offering to someone who has failed. But instead, that we're there together as people who are trying to find a way forward, and at this moment you may happen to be the patient, and I'm not. But our connection is going to be enriched if we both recognize that at some point, I may be the patient—and that at some point the person who is the patient may become the person who is the helper.

It could be the case that our wounding comes from a chronic physical condition, which allows us to have empathy for and insight into the truncated lives our patients often live. The wounding could be a difficult family situation that allows us to empathize with a family that has a hard time showing up in the intensive care unit (ICU) because of the complexity of their lives. Anything in our life can become a teacher for us and an avenue for empathy and connectedness and presence to another person.

This is something that I talk to resident physicians about when doing debriefings in ICUs. We'll be discussing a difficult case that they have, and I will say:

> Every clinician that I've ever talked to, at whatever age—85, 75, 45, 35—has at least one case or one patient who stays with them always. Don't consider that a failure. What that patient is, what that case is, is a teacher.

A teacher? By that I mean that this is a person or a situation that you will continue to learn from, that will continue to be a resource and a touchstone for you going forward.

Allow me to give an autobiographical example because I think it sheds light on this enigma of how wounds can become teachers, can become resources. I've been a hospice volunteer for more than 30 years. I've spent the past 10 years working in ICUs in both adult and children's hospitals in two different academic medical centers and during that time taught a course on death and dying for undergraduates for several years. The core resource for me throughout my work has been the death of my father when I was 18 years old. Not just the death, but the way in which it occurred, changed nearly everything about my life. It remains at the core of who I am today, even though it happened more than 50 years ago.

In a conversation a month ago, a patient's family member said to me, "I can hear the pain in your voice when you talk about how your father died." This came forward in such a way that it was not a distraction from what was going on in the meeting. It wasn't about neediness or therapy. She wasn't saying, "Oh gee, this guy who's supposed to be helping us is stuck with his father 50 years ago." Instead, it was this woman acknowledging, "Yes, you feel the same pain that we're feeling right now." It was recognition and connection. The presence of that wounding has become a resource, not only for me but also for the people with whom I work.

Let me explain that in a little more detail. My father was 43 years old, a little bit overweight, had some high blood pressure but basically no health problems. I was 18 years old, newly graduated from high school. Although we had fought about this, that, and the other, I loved him. I was opposed to the Vietnam War; he was keen on seeing me over there in combat. We quarreled about many of the political issues of the day. But I adored him. I adored the ground he walked on. He was the central person in my life.

One Saturday afternoon in July, he went out and played golf and came home with a bad headache. He'd had headaches all his life, migraines. Dad with a headache wasn't anything unusual, but this one somehow seemed a little different to me. I stayed in that night. It was a Saturday night, and I normally would have gone out with my girlfriend. But I stayed home, thinking I might be needed. Early in the evening, I heard my mother screaming upstairs. I ran up there to find my father's right arm and leg jerking all around. The leg was almost at a right angle up from the bed. Convulsions, seizures.

We called an ambulance, and while we were waiting for it to come, I moved him up on his side so I could rake the vomit out of his throat to prevent him from choking. He was soaked with sweat, in horrible pain, unable to communicate. He finally became completely unresponsive, comatose. The ambulance took him to the hospital. He never regained consciousness.

He had had a huge stroke. We were told that there was a ballooning in the ring of arteries at the base of his brain. The doctors called it a berry aneurysm and said it had exploded and basically wiped out his brain. The next morning, they did a surgery, which was certain to be futile, but it was done anyhow. The events that unfolded after the surgery have informed my interactions with patients and families ever since.

My mother and I were in the waiting room. The lead surgeon came in and motioned for us to come out into the hallway. Standing there with us he said, "This is one of the worst strokes I've ever seen in my life. He's going to be a complete vegetable; nothing we can do for him. You need to decide whether to take him off life support, or not." And he walked away, leaving us standing there in the hall. My mother was 42; I was 18. Fortunately, there was another surgeon there who knew us, who was able to find us a tiny room where we could sit and talk things over. We didn't talk for long; I don't think it was more than 15 or 20 minutes. It was clear to me, even at 18, that he was dead, or essentially dead. And the idea of this vibrant, active man being on a breathing machine in a persistent vegetative state was repugnant to my mother and to me. We decided in short order to take him off life support.

To describe the whole range of changes in my life that these events brought about would not be to the point. But I will say that at that point, my life was fundamentally recast. I was on my way to college, intending to major in political science, go to law school, go into politics and work on public policy. Instead, I ended up majoring in religious studies, comparative religion, and philosophy. I needed to try to begin to understand this life, this world, in which events like this could happen to a good man, to a person I loved and cared for very deeply.

Many years later, I was sitting with a hospice patient, a homeless man who had been released from prison because he had AIDS. He was decidedly skeptical of this planned visit. He had seen a lot of do-gooders. This was a multiple felon who'd been released from the state penitentiary because they didn't want to deal with his AIDS configuration. They didn't want him sickening and dying there, so they turned him loose. I go in and sit down, and he looks at me and he says, "Mr. Shanks, what are you doing here?" Absolute scorn in his voice. And I'm thinking to myself, "Okay, I've got one chance, because this guy is determined to find bullshit." I decided to go for broke. "Mr. Jones," I said,

> when I was 18 years old, my father died in my arms. I'm 43 now and I've been trying to understand death ever since. You're dying,

and I'm here to learn about that—to learn about your death, and to learn about death from you.

I don't think anybody had ever said anything quite like that to him before. I had given him as honest an answer as I could because that *was* fundamentally why I was there—and he could hear the honesty.

He sent me out to get a six-pack of cream soda, of all things, and after that we were buddies. And I was with him the night he died, about a year later.

Another example: I was working with a family, the father of which had been in a motorcycle accident and had a terrible head injury. He wasn't brain dead, but in a persistent vegetative state. This man had a son who was 24 years old. They had just opened a business together. The patient was divorced. The decision-maker for whether to withdraw life support or not was therefore this 24-year-old son.

For obvious reasons, the son didn't want to come into the hospital. He didn't want to see his dad in the trauma unit. And he knew that people were going to ask him to decide to remove life support on his father, and he didn't want to do that. He kept not coming in, and not coming in, and not coming in. I finally called him, and I said, "Johnny, I want to tell you something about my own life. I want you to see that when I say I understand something about this, I actually do." I told him the brief version of what had happened with my father, and the decision that my mother and I had made about the life support. His first response was, "God, you were young." All of five years difference. But he was genuinely moved. I'm thinking of him as 24 and a kid. He's like, "Oh my God, 18. You were young." This brief interaction did two things for him. One, it drew an element of compassion that allowed him to jump out of this horrible situation he was in. His dad has just been in this motorcycle wreck, and yet here he is moved with compassion for my 18-year-old self of 40 years ago. Two, it established a connection between the two of us. He came to the hospital two days later for a care conference with the team. At the decisive moment in the meeting, he said, "It's time for us to take my dad off life support."

When the meeting was over, I leaned across the table, shook his hand, and said, "You're brave. You're very brave. I'm very, very proud of you. I know how hard this is." We had this totally direct connection, which we never would have had without my father dying the way he did, and without my having spent decades learning from what had happened. The critical thing is to recognize and treat the wound. This was more than a single wound for me, it was a series of them, a total catastrophe. I was gutted to the core. But gradually, across the decades, I learned to treat that awful series of events as a resource and a teacher.

II. Working Toward Healing: Inner and Outer

One of the most important questions is how to go to work at the point that we recognize our woundedness. In the first three phases of the healing vocation—call, formation, and mastery—the focus tends to be on the vulnerability of the other and the woundedness of the world. But in the second three phases—depletion, breakthrough, and replenishment—one's own woundedness necessarily plays a central role. There is the iteration, the reverberation, between one's own woundedness and the suffering of others that I am calling moral anguish. This anguish is certainly a driver in depletion, but—perhaps counterintuitively—in breakthrough and replenishment as well. Recognizing and addressing anguish will be one of the most critical steps along this path of the healing vocation.

A recurring challenge, then, in each of the six phases of the healing vocation is how woundedness is held:

1. Call—to address world's wounds
2. Formation—preparing for the encounter with woundedness
3. Mastery—success in healing wounds of others

4. Depletion and burnout—one's own wounds come forward
5. Breakthrough—one's own wounds, and the wounds of others, as teacher
6. Replenishment and renewal—one's own wounds, and the wounds of others, as resource

We are simultaneously healing our wounds and being crippled by them, taught by them, fed by them, at every step along the way. Which raises for us, in turn, ongoing questions about vulnerability, wounding, and healing—how they're related; how they're balanced; which side dominates; how both can be held.

It can be useful when working with one's own woundedness to distinguish, in a rough-and-ready way, three broad types of wounds, and primary strategies for approaching them. The most straightforward is *present-time wounding*: events and situations that are distressing and injurious, which are occurring now or have occurred in the past few months in our work setting. Second, there is what can be called *historical wounding*: events that have occurred at some past point in our careers or adult lives that continue to shape our daily experiences. And, finally, *archaic wounding*: events that occurred in our childhood—including physical, sexual, and emotional abuse, and other trauma of a similar degree—which took place well before one's career formation. Addressing archaic wounding becomes steadily more important as one's career develops, for it does seem that extensive early wounding increases the risk of full burnout.

Typical sources of present-time and historical wounding in healthcare include

- patients who challenge us (especially the violent and abusive);
- pandemics (AIDS, COVID-19);
- mass shootings;
- natural disasters (Memorial Hospital during Katrina); and
- civil war and international war.

Even as one is being overwhelmed by present-time events such as those listed, it can still be helpful to recall historical wounding. For instance, reviewing clinician challenges faced during the AIDS epidemic has proven useful to many challenged by the COVID-19 pandemic—the institutional failures, the political tangles, and the startling disease presentation all being common themes. On the other hand, during present-time events of such scope, it is almost impossible to provide the safety and resources needed for the in-depth work addressing archaic wounding requires.

For coming to terms directly with present-time wounding, regular debriefings with senior colleagues and ongoing peer support are helpful. As is the pursuit of advocacy opportunities at every level: personal and institutional; political and social; local, regional, and national.

Exploring historical wounding is generally best done with the assistance of another person or in a facilitated group setting. This wounding is most often rooted in highly emotional and unresolved conflicts. Having another person present—preferably one with professional training—to assure safety, provide guidance, and offer reassurance is invaluable.

For engaging archaic wounding, being accompanied is absolutely essential. Let me say it again: It should not be done alone. Seek assistance from a therapist with expertise in this area, a counselor in your workplace, a priest or minister, an elder in your community, or an older colleague you trust. Whomever you choose must be able to provide and maintain a safe, stable space—emotionally and physically—if healing, not re-wounding, is to take place.

Having done some of the exploration encouraged above, we can now go further. Finding connections and continuities between the suffering outside of us, which we have chosen to address, and the wounding inside us can be revelatory. Why have we chosen bedside nursing? Why mental health? Why domestic violence? Are there experiences earlier in our lives that have impelled us to take up this specific cohort of patients, to go into this specific work setting? Can you see links between what drives you as a caregiver and your historical or archaic wounding? Why are your energies and your gifts and your abilities mobilized by being in the presence of one type of suffering rather than another? Again, are these dynamics related to earlier wounding? The answers to these questions in no way trivialize the vocation or explain away motivations. They simply provide more information and more insight, which in turn increase our capacity to do healing work.

We feel a call to do a specific type of work, and we open ourselves to it. We open our hearts to the suffering and challenge and pain that's around us, pain that's commonly beyond what is served up in our own lives. And we step forward to do that willingly. What is that willingness about? One of the elements of it is certainly compassion. Another is the desire to be connected to those who suffer and to feel the power available in that connection.

To go back to Mr. Jones, I was with him the night that he died. I was a relatively new hospice volunteer and didn't understand enough about what the resources and protocols were. More important, I did not know what his dying ought to look like. He had an especially vicious tuberculosis, common at the time in AIDS patients. Basically, he died drowning in his own fluids. We weren't in a hospital unit. We were in a house for homeless people with AIDS. Some of his family members volunteered to go get the hospice medications. It didn't occur to me that because the medications were morphine and Ativan, they might pick up the prescription and take off. An hour or so later, I realized they'd disappeared with the drugs we needed to address his pain, his anxiety, and his oxygen hunger.

If I had known then what I know now, I would have called hospice and the hospital, and raised hell and demanded that somebody come. But I didn't get it. He had pain; he had fierce oxygen hunger; he screamed on and off for hours. It was a horrible, horrible death. True, it would have been worse if he had been completely alone, and possibly worse if he had still been in prison. He had several siblings who lived within a few blocks of where we were that night. Most of them were in the living room, but only one sister would come in the room where the patient was. There was such fear of AIDS then. That sister did help us change his diapers but wore gloves the whole time she was in the room. The only person who touched Mr. Jones skin-to-skin that night was me.

I sat with him. I had my head up against his ear talking to him for hours. An agony I'll never forget. And if I had known what I should have known, it wouldn't have gone down that way. There's no way I would let that happen now with anyone. But I didn't know enough. That's okay; he continues to teach me. It's an example of a death that shouldn't have unfolded the way it did, and for which I feel direct responsibility. But he is behind teaching opportunities with residents, with undergraduates. He's behind my caring for homeless people dying of AIDS for over a decade. He has sponsored an enormous amount of growth, which has resulted in resources I've been able to offer to hundreds of people since then. He left a profound legacy in my life and the lives of many more beyond me.

Everybody who works with those who suffer finds ways to learn from their darkest moments. I'm not special in this regard. These are the examples I know best and am freest to talk about. The truth is that if we can learn to hold those most vulnerable and broken pieces of ourselves and our lives gently, without forcing them away, without being torn apart by them, to hold them kindly, we're much more likely to be able to also hold the brokenness of others, and not be debilitated by the fact that we can't fix their brokenness. Not be frustrated or enraged by the fact that we can't fix it, but able to allow ourselves to stay put and to be present to it. Because we know from our own lives that there are these pieces that are never going to be different. My father is never going to die in a different way; that's permanent. But how such facts are picked up, and what we learn to do with our pain, or indeed more likely, what it does to us as we move forward—there is an enormous range there. One thing that potentially can happen with the breakthrough process is that we can be set free from our stories, and the sets of expectations derived from them, which have hemmed us in or narrowed our perspective up to now. We have the chance to get a broader view of what our own lives are about, as well as the lives of the people that we are working with, talking to, learning from. With breakthrough, we can hear the call to be a healer in a whole new way, on a more impersonal level, with our hearts revived and expanded.

III. Openings for Breakthrough

Talk about depletion, and "burning off" or "burning out," focuses reasonably enough on how difficult all this can be. But we need also to recognize that these experiences can be salutary if they become catalysts for further growth and moving into greater capacity. There is, in short, hope for what we are calling breakthrough. It's good to be careful about this word, because it's not like we break through once, and we're home free. What we are talking about is a breakthrough into a larger perspective, and into new sources of energy and inspiration. Into a bigger picture such that one can be present to a wider range of experiences, a wider range of suffering, without going through the process of deflation or discouragement. The hope is that once you reach a point where things are so painful that you can't continue with the old ways, you are driven or forced to reexamine yourself, your skill set, your work, your workplace—to reexamine what you're doing, who it is that you work with, and where you work. Pain is a great motivator for change; it's a process that can catalyze transformation.

To engage one's depletion in such a way that it becomes breakthrough rather than burnout, it is essential first to observe what is going on. We're not going to be able to imagine the scope of possible transformation if we can't see the scope of the pain. This is a simple point. But the reality is that not looking even as we suffer is itself a way of dealing with adversity, stress, and anguish—in other words, numbing oneself, or avoidance. Not registering what we know we cannot handle, even as we are in the middle of dealing with it—that unending hour-by-hour stream of cases in the trauma unit, for example.

Details are important. Saying, "I'm exhausted, I need a vacation," doesn't get us very far. Go for more granularity. Describe a dozen interactions with patients or colleagues that are exhausting. What specifically is exhausting about them? And boredom: What does that feel like in your body? Where do you feel it? When do you feel it? Is it truly boredom—or exhaustion—or feeling trapped?

And the patients—what are their most common conditions: "I'm tired of these ear infections. I'm horrified with these rat bites. I see these same injuries over and over." What is it about your day that's frustrating? For some people, it would be, "Well, if the parents—or the schools—would do X, Y, and Z, these kids wouldn't get these infections, these bites, these injuries." Or it might be, "That's all that happens in Vermont. The only problem that anybody comes in with is an ear infection." We can't let ourselves off with, "I'm bored" or "I'm exhausted."

In choosing a healing vocation, we are committing ourselves to—or throwing ourselves toward—specific processes of transformation. I'm giving here one picture of what those processes might look like. The suggestion is that coming into periods of depletion and discouragement is to be expected in this developmental process and ought not be taken up as failure.

Entering the field of suffering, we ourselves are in play, alongside our patients. We're not static; we're in motion. Pressures and demands are working on us. And the more we can enter into these processes consciously and as an active participant, the more transformative, power releasing, and life giving they can be. The key is learning to cooperate with those forces that are impinging on us. Impinging on us in present time, in future time, in past time—which all come together in the situation we're thrown into, or are throwing ourselves into, or are being thrown into.

I like to say: *The hope is always in the particulars.* Interestingly enough, we tend to think of abstractions as having to do with ideas. But emotional experiences can be abstractions. "I'm tired and I'm bored" is an abstraction. It's a trap. Because there's not enough specific information to give any idea of what a new strategy might be, or what a way out might be, or what the actual situation is. The shorthand of "I'm tired and I'm bored" is, in effect, most commonly, a muted translation of "I'm stuck, and I like it." My experience is also when people are encouraged to, and given the chance to, speak about the details, there is almost always an enlivening, even if the description in the details is horribly painful.

Just this basic enumeration has the potential to overpower simplistic narration. Once you begin to describe the cohort of patients you have, it's almost impossible not to realize, for instance, that there are lots of other kinds of patients. Some of whom you might rather treat than the ones you have; some, definitely not.

None of us are infinitely plastic. It's simply who we are; there are limitations to us all. But if we don't look at the particulars, we can't see, we don't see, we won't see. We live in the particular; our enlivenment is in the particulars. When we go abstract, either emotionally or intellectually, we are evading what's going on. The power that is there in the thickness of what is happening is where the energy will come from to move us to the next place.

The point being that one of the dangers in burning out is getting stuck in a negativity shorthand loop. We don't look at the particulars. We just say, "I'm bored and I'm tired." And then we are stuck; there is no way forward. So, the first thing is to look at our lives in their actual detail. That, in and of itself, is going to break through some of the numbness. It's likely to be painful—and that's good! *Because the real danger with depletion and burnout is not the pain, but the numbness.* It begins to feel like the only solution is to refuse to feel anything anymore—good, bad, indifferent.

The other key to making it to breakthrough is taking the developmental view. And for this, having a map is helpful—a map that allows us to expect that there will be something on the other side of the pain and the negativity. This is one of the primary reasons I introduced the typology of the six phases. Because one of the difficulties, once more, is the way burnout is depicted as a dead end. "You're a failure if you've ended up here." But if these movements can be seen as a piece of a developmental process that is to a degree inevitable, we're no longer in a dead end, we're just in a thicket. A dark wood, if you will. Tangled up in vines, maybe, and knee-deep in the muck—but at least there's a way out.

IV. The Great and Unexpected Depth of Vocation

At this point you may well be asking yourself, "Why do we have to go through this? Why is it part of us?"

To which I say, "Yes, I know! What a question! What a life!"

As you were drawn to the healing vocation, you stepped forward because you had gifts to offer. You also had sensitivities and awareness of suffering, and a willingness to deal with difficult things. Even if you say you're in it for the money, you learn the diagnostics, you smell the smells, you work the hours, you do whatever it takes. But at some point, I would argue, it will be useful for you to ask questions about why you have this sensitivity and this willingness in the first place. Why can't you look away? Answers here may provide the first clues about historical or archaic wounding that have been forgotten, or repressed, or otherwise locked away in the past, locked away in the body.

Yet, always, "Your pain will find you out." At a certain point, for instance, you might discover that you have a distinctive breathing pattern that is a wound that is written into your body. You may not get to it through a psychotherapeutic history and exploration of family. You may get to it in terms of, "I hyperventilate in situations like this." Or: "I have a hard time getting my breath," or, "I feel like

I'm suffocating." You may end up coming into your woundedness as it manifests in your body—that may be your access to it.

You might discover, for instance, that the best way for you to get past these bodily disruptions is to do a course of psychotherapy addressing early trauma. On the other hand, you may be a person who does an enormous amount of psychotherapy, and yet reach a point where your therapist says to you, "That's as far as we can go with talk therapy. You need to go do breathwork. You need to go do bodywork. You need to try craniosacral therapy or deep-tissue massage." There are different paths into the wounding. Get at it through whatever draws you in and sustains you.

These various examples point to the fact that these processes aren't ever going to be easy, nor will you be done with them quickly. Often they may be quite drawn out, and you may have to endure repeated cycles of counseling or long periods of depletion. How to sustain ourselves through it all? The key point here, which is admittedly harsh, is that *this process is your life*. Which is another reason to go for the particulars. It's not like we have the option of saying, "Okay, I have my life. I'm going to do this process in relation to my work." The teacher Cheri Huber uses this wonderful saying: "How you do anything is how you do everything" (Huber 2018). There are certain core patterns that show up. They replicate themselves in various places in our lives.

Which means that one answer to that question of "how" is: "We don't have any choice, because this is our life." If we don't do that hard work in this setting, we're going to end up being faced with it somewhere else. If we don't find and address issues that come forward in our workplace and are leading us to depletion and burnout, and deal with them there, they're going to erupt in our parenting, or in our intimate relationships, or in our congregation, or on our softball team. That material, which is who we are, is going to express itself somewhere. We don't have anywhere to run; the only option is to refuse to engage our lives. Which refusal is what our addictions are about—whether we're addicted to substances, or consumerism, or sheer activity. This is what our wide variety of evasions are about. One of the most critical realizations on the way to breakthrough is that, finally, evasion is not an option, nor is escape.

This is another place where it's useful to step back and recall what we know from basic biology and physics about homeostasis. It can be disrupted by unbeckoned events, and it can, in turn, recover into balance. This process of ebb and flow is something that we know about healing, as well. Being alive means being involved in the process of going through periods of disharmony and imbalance, and coming back into harmony and back into balance—going through cycles of wellness and illness. That's true for our patients; and it's true for us.

Those transitions in and out of balance, in and out of harmony, can be quite painful, and yet are utterly inevitable. Painful because they occur on so many different levels, and all at once—and each level has its own rate of change, not necessarily coordinated well with that rate on other levels. Our bodies move at one pace, our families at another, our work world at another.

And there is the exceptional nature of the healing vocation. If we look at a book such as *The Body Keeps the Score* or any of Peter Levine's works on trauma and mammals, we learn that there is a complex, automatic set of physiological responses to trauma built into our bodies (Van der Kolk 2015; Levine 1997, 2010). For many, if not most lives, these in-place safeguards supply all the protection that is needed. But there are lives that are so compounded with trauma that the normal mechanisms aren't sufficient. That's when we see enormous brokenness: people we see living on the street, people we see constantly returning to our emergency departments and psychiatric units. We see them in addiction centers, and we see them in prisons, these lives where the compounding of trauma overwhelms the natural defenses.

For the most part, though, the mechanisms developed across ages of evolution tend to work. But things are different if our calling, if our given work, entails taking on additional amounts of suffering.

If, in addition to the cup and a half of normal suffering that gets poured into any life, I'm willing to go to work every day and stand in the field of suffering of person after person, after person, then I have a responsibility to those people, and to myself, to develop ways of dealing with that excess burden.

In conclusion, what I want to say is this: Depletion is likely inevitable. Breakthrough is possible. Burning off can be survived, and even profited from. Burnout itself can be avoided. It can also be survived; recovery can happen. There are innumerable ways to live with and through these processes: staying in the job and finding different ways of working; reconfiguring the job; finding a new iteration of the work in a new location; leaving this work for another line of work altogether. We must be true to the particularities of our calling. We can break through to replenishment and renewal.

4

The Practice of Replenishment and Renewal

Core Exercises

The body itself is our moral presence, and our most essential healing resource. The body itself is, additionally, where moral anguish is experienced as assault and as injury. This in turns means that an essential component of breakthrough is developing skills to support the body in its healing work—the healing of others, and the healing of itself.

Breakthrough can thus be thought of as a new formation process leading to new mastery—a re-formation, leading from depletion and burnout to replenishment and renewal. The meta-skills approached in the exercises in this chapter are, in short, a set of opportunities to learn to abide in the field of suffering and transform it into a field of healing.

In this chapter, I'm going to be talking about exercises I use regularly in workshops related to vocation, burnout, and moral distress. In Section 1, I'm going to describe how I learned the two foundational processes, because all the exercises outlined here are essentially variants of awareness practice. In Section 2, I will go on to describe other core skills focused on awareness of interactions. Finally, in the book's appendix, "A Practice Calendar," I offer a concrete program for integrating many of these skills into daily work life.

One thing that distinguishes my approach from the common teachings on mindfulness and stress, with which you may already

Into the Field of Suffering. David Schenck with Scott Neely, Oxford University Press.
 DOI: 10.1093/oso/9780197666739.003.0005

be familiar, is the focus on moral tension, and on the specifics of healthcare settings. My conviction is that listening to the wisdom of the body's distress is the most essential skill of all.

Unlike most discussions of mindfulness exercises in healthcare settings, I place far more emphasis on the body and its states than on mental and psychological features. That means these exercises should be approached as being akin to exercises you might be assigned by a physical therapist. We're not aiming for the quiet of a yoga studio, or the contemplation of a monastery or zendo. We're not aiming at establishing a mindfulness practice. What we are looking for is much more focused and immediately practical. We need intervention skills with which to disrupt trauma and stress formation. We are looking to short-circuit wounding, and to turn experiences of rupture into ones of bodily awareness and healing (Ray 2016; W. Johnson 2000).

Section 1: The Foundation (Internal Awareness)

Awareness and Physical Pain

Let me begin with a personal example. My first experience of the effectiveness of awareness in dealing with physical pain came during a meditation retreat run by the Zen teacher Cheri Huber (for insights into her work, see Huber 2000). Cheri's retreats at that time were set up with a one- or two-hour work meditation period in the mornings and the afternoons, with four to six 30-minute sessions of sitting meditation, and one or two dharma talks by Cheri.

Consequently, one spent a lot of time in meditation posture on these retreats—hours and hours a day. For beginners like me, that's a huge initial challenge. And on this retreat, the place where we were staying had a major building that had recently burned down. A central task for work meditation for many of us was clearing this site. The ground was mostly red clay, and there'd been lots and lots of heavy rain that had soaked into it. The digging and the lifting were heavy-duty work. At some point in one of the dharma talks, Cheri talked about working with physical pain during the meditation. When I went into my next private session with her, I said, "My back is killing me from all this shoveling clay and pushing wheelbarrows. What do I do? How can I work with this?" I had heard that Cheri had injured her back some time ago. And of course, since she sits in meditation all the time, I figured she would have to have something useful to say about back pain. The main thing she said was, "Pay attention to what's going on."

Meaning: Try to set aside your narrative about it and pay attention to the pain itself. Set aside the verbal label "pain," and pay attention to the sensations. Try to go with the sensation alone, looking for its epicenter. Go to the hurting place in your body, listen to it, affirm it. The idea is that the pain is a piece of your body screaming for attention. And that if you rest your attention there, you will have a different experience with what's going on there. Another thing she said that was very interesting was, "Try to catch it before it turns into pain." As soon as you sit down in your meditation posture, you know it's going to be hurting in two minutes or five minutes—or 30 seconds! But take your attention immediately to where you expect the pain to come and see if you can catch it when it's just a tickle, just the first little inkling of, "Ah, my back is going to hurt." See if paying attention to it changes the experience. And to my amazement, it was remarkably effective in short-circuiting or buffering or shutting off the experience of pain.

With regularity Cheri would also talk about working with emotional pain in a similar way. "Where do you feel this emotional distress in your body?" In the Buddhist understanding, emotions are labels put on sensations, and sets of sensations. Which means you should ask yourself, in working with emotional pain: "Where are the sensations in my body?" Okay. It's in your throat, where exactly in your throat? Look for a very specific point, asking yourself what the epicenter of the bodily sensation is that goes with your emotion.

Rest in that place with acceptance of what's there. Try to go into it and feel it and see how it's transformed by your attention. All this led me to think, "Surely if I can change my experience of sharp physical pain in my injured back, this technique will also work with emotional pain."

In the process of developing these exercises to use in workshops, what I have encouraged people to do is to try to be aware of whatever physical pain they have associated with moral anguish, depletion, and burnout, and all the distressing emotions around them. I say: "Try to find where they sit in your body. Try to feel your way into that and see what happens when you give the distress direct and explicit attention." And then I encourage people to expand that awareness by trying to see and feel when and where the pain first begins to move, and to notice which kinds of work settings generate pain, and which kinds of interactions, and with whom.

Being in the emergency room may make your throat tight, while being in the clinic makes your back hurt. Or it may be a specific patient causes you pain in your shoulder, or a patient with a different condition causes you pain in your neck. Try gradually to become more aware of the flow of care's labor through your body.

As I've emphasized, the work we do with those who suffer involves embodied presence. It involves being in the quite literal physical field of people who are in pain of one kind or another. This is a field charged as well with intense emotion. And all this is going to flow into your body. Which means that being able to operate in that field as a healer, and without having continually to remove ourselves from it, is going to require learning how to monitor our bodies and understand what's happening in them.

Pause–Center–Shift

The next exercise comes through work with a different teacher, Brugh Joy. Brugh was a physician who recovered from pancreatic cancer, became a spiritual teacher, and taught, among many other things, heart-centered meditation. Brugh used to talk about a process he called "pause–center–shift." The idea behind the practice is that, when you feel yourself on the verge of being overwhelmed, when you feel negativity or reactivity rising in you, you stop whatever it is that you're trying to do, and step back a moment from whatever is going on, inside or outside of you. Mark that *pause* with a long exhale and a couple of deep belly in-breaths, where you expand your abdomen rather than your chest. Then come back into your *center* and seek balance. Perhaps you've been doing heart-centered meditation, and you bring your attention back to your heart space. Maybe you have a sitting meditation practice, and you deliberately shift your posture to get your spine aligned and centered and straight. Or maybe you have a centering prayer practice, and you take this time to repeat your prayer word or mantra a few times to bring yourself to center. Whatever it takes, however you can manage it, come back into a balanced and stable mental, emotional, and physical state. So, that's *pause* and *center*.

Shift is a bit more complicated. Brugh held the position that we all have multiple personalities (Joy 1990, 2010). One of his favorite things was to describe how different personalities could have different physical symptoms for diseases, but obviously still be "inhabiting" the same body. Fascinating stuff that he was interested in. But, on a more rudimentary level, his idea was the not uncommon one that we have multiple selves, or at least a self with multiple aspects. He urged us to learn to "change hats," as it were—to practice shifting from one aspect or personality to another. Pause, come into center, and then shift out of the part of you that is young and scared by a situation you're in, and go into another part of yourself that is not afraid, that is more oriented or maybe even wise (Joy 2008)!

Pause, center, and then move to a different set of capacities in yourself. For a clinician, it might go like this: Pause this conversation where I'm trying to explain this medical procedure, its pros and cons; center, and then shift into the part of myself that is fundamentally present and empathetic, and do nothing but listen. The critical part, which matches nicely with all awareness and mindfulness teaching, is that to initiate the pause–center–shift process, you must be aware of how badly things are going in your body and with your emotions.

And the thing is, we often don't realize how badly things are going for us until the train's already wrecked. But with an awareness practice, you can begin to learn things like, "Once I get this tightening in my right shoulder, I know that bad things are probably going to start moving for me, and the panic attack will begin." Or the nausea, or the dizziness, or the migraine—however it is that your body works to get you to pay better attention to what's going on around you. And you will know this because you've been paying attention to your body and to how negativity flows through it. At the point you realize your own stress process has begun, you do the pause–center–shift practice. (Cf. Rushton 2009.)

Section 2: Expanding the Frame (Awareness of Interactions)

Listening

The critical interactional exercise is mindful listening. In workshops, I have people work together in pairs, and instruct them to follow carefully a specific structure in their interchanges that is designed to prevent conversation back and forth between the partners. One person has five minutes to talk about the assigned question without being interrupted or prompted (e.g., "I first knew I wanted to go into medicine when . . ."). Then the listener has two minutes to summarize and validate what they've heard. And then the original speaker has one minute to validate, or to correct. To say, "You totally missed the point!" or "Great, that helps me see 'X' differently." Then the roles are reversed, and the pair goes through the exact same procedure with the exact same prompt.

When you do this with groups, you discover it's very difficult for most people not to turn the exchange into a conversation, with its habitual back-and-forth. But one of the things we know is that an utterly critical skill for building relationships with patients is listening. We need to learn to listen, deeply and carefully. And that takes practice.

The other thing I always say when doing this exercise with healthcare providers is, "Here's your opportunity to be listened to. You've got five minutes to talk, and somebody's going to be a devoted listener for you for five whole minutes." In general, nurses, physicians, therapists, and caregivers *never* get listened to. In these roles, we're doing the listening. Here instead is an opportunity to be the speaker, and to be totally listened to by somebody who's totally focused on your stuff, and not swimming in their own, as most of your patients and clients necessarily are. The core structure—the clarity of the roles, and the swapping places—is an essential part of how the exercise does its work.

Hot Spots/Guided Imagery

Another thing to do is to use guided imagery to explore experiences and feelings that have been pushed away: "What was it that I wanted to say that I didn't? Is it something that I need to say, or not?" There may be personal archeology to be done, the exploration of your early life. There's a good chance, for instance, that at an earlier time in your life, there were things that you wanted to say but didn't, and now a situation you're in echoes or reverberates with that unfinished business. You could potentially build out from that, too.

This unfinished emotional business can reside in your body as hot spots. Remember stubbing your toe as a kid. And bumping it again afterward, and how painful that was. Emotionally, a similar process can occur, and can operate across decades. We're sensitive about certain kinds of comments or actions; we overreact. The point is to locate these sensitivities in your body and try to heal them before they get bumped again. Or to at least reduce the pain from the bump!

To do this, get into the relaxed posture, exhale fully, breathe in by expanding the belly, and bring forward internal kindness, for the places in your body that are tense, overheated, painful. Do this systematically, day by day. Go through each day and think about the kinds of patients you've seen, or the kinds of clinics you've worked, or the units. What is it in those different encounters that makes you most uncomfortable? Where is that discomfort in your body? Gradually develop a map of what's going on in real time. This map can be used to help you explore your past; it can take on an archeological dimension. Such mapping has the potential to transform the way your work experiences unfold. Work with your map and gradually allow each of these knots to dissipate. Often you will have memories associated with these spots and these emotions. Allow them to come forward; accept them; watch them move on. If you do this often enough, you can modify the very immediate reactions you have in the lived present and improve your ability to sit comfortably with those reactions. And, with diligence, you can learn to catch your own reactivity before it becomes full-blown.

Mirroring and Breaking the Circuit

Another related practice is to become aware of the ways in which our bodies mirror, or don't mirror, the body positions of the people that we are talking with. Or we may find that a person that we're talking to is imitating the posture we are in! They may be tugging on their ear when we do; or they may have their chin in their hands when we do; or they may be clenching their fist when we do. There's interesting research on this as a biological level of mimicking, which is hypothesized to have evolved to help generate emotional connection.

But if the emotions that are moving in this connection are negative emotions, you don't want to be trapped in a closed loop with them. In challenging encounters, ask yourself: "Is there mirroring going on here?" And then: "How can I change my posture so my body is not emotionally in sync with the person where there's negativity or reactivity going on?" This is another place where coming into a neutral posture by straightening your spine and planting your feet can be helpful. It may keep you, for example, from either leaning forward into a more aggressive posture than you want or leaning backward into a more passive or receptive posture than desired. It can allow you to stay neutral and be a safe and calming point in a tense room, not moving one way or the other, not expressing one thing or another. "Holding the pole," I call it.

To me, this is all still awareness practice. It begins with trying to pay attention to what's going on in your body and watching the reactions your mind automatically produces. Once you have developed the ability to get still and to pay attention, you can apply these skills anywhere, anytime.

Think about different patients, think about different procedures that you do, different conversations you must have; think about different physical settings, hallways, this room, that room. Maybe your unit doesn't have rooms, maybe just curtains between patient beds, which has a whole different level of stress to it. Or let's say you're working in a community and there are certain streets or certain neighborhoods or certain blocks that you know are going to generate certain kinds of emotions. Think about them, how you use them, how you approach them. This is a level of attention that we don't normally give the details of our lives. What that means is that our body is in the process of gathering and storing all this information—and all this accumulated data and our reactions to it are having an enormous impact. Of which we're mostly unaware. But if we are paying the right kind of attention, we can improve our ability to move with and assimilate the massive inflows of energy and information that are always occurring, especially in stressful and threatening situations (Rothschild 2006; Gallese 2003; Hatfield, Cacioppo, and Rapson 1994).

From Empathy to Presence

I've talked about shifting from reactivity to presence. Now I want to talk about another shift—the shift from empathy to presence. Suppose you're experiencing negativity in the room around you, or you're experiencing reactivity in yourself in response to that negativity. You do the pause-center-shift movement; you come into a settled posture; you open yourself further to what's going on in the room. At that point one good thing that can happen is the arising of deep empathy for the people who are in the room.

Empathy of course is incredibly important when we are with people who are suffering. Empathy, we may say, is being able to imagine yourself in another person's position, and not just have sympathy for them, but reverberate with ideas and feelings like the ones they are having. It's an important capacity, but a potentially disabling one as well. (Cf. Sobel 2008.)

We often think of empathy as the root of compassion. We tend to assume that the way to hold a compassionate position is to stay in an empathetic position. What I want to suggest is that there may well be times that empathy itself is something that needs to be metered. There will be times you need to step back from the empathetic stance. This does not necessarily mean literally, physically stepping away from the person you're with—though that may be appropriate in certain extreme circumstances. But it does mean, for example, coming regularly back to center, reaffirming your posture, reaffirming your breath, and bringing your emotional bubble or sphere of influence back into your own body. Again, this doesn't mean you depart from the person who's in front of you. It doesn't mean you stop listening to them. It doesn't mean that you stop responding to or caring for them. But it does mean that your emotional center shifts from being directly connected to theirs and comes back into your body, into your center—the heart, the spine, or the central channel. You continue to be present, but the emotional circuits are running differently. And it may be that, after you have come back to center and held there for a few moments, you are able to move back into an empathetic position. If you monitor yourself carefully, you can move back and forth safely, and without losing touch with the other person. (Cf. Halifax 2018.)

True, empathy is an important skill, but like all skills, it has limitations. There are times when it's important to set it aside. But also realize that that doesn't mean that you're no longer being compassionate. This is an argument that presence is the core of compassion. And that compassion may involve, in certain situations, an intellectual or analytical response, rather than an empathetic one. Often of course compassion does involve an empathetic connection and allowing people to see that you're feeling what they're feeling, and to see that you know what their range of distress is like. But there will be other times when that's not called for.

Back to what I was saying earlier. All this builds on the awareness practice; it's a further, more advanced step. And what we don't want to see is the evaporation of empathy: "I'm not going to get caught out like that ever again." How can we work with that inclination to shut down, to withdraw? Maybe, for instance, I am in an empathetic stance too often. But that doesn't have to mean, "Don't be empathetic." Is there an alternative between empathy and disconnection? What I'm suggesting is that, yes, there is such an alternative. There is a way to remain centered and self-contained emotionally, while your attention and your care remain with the person who is in front of you.

5

In Conclusion

On Healing Presence and Gratitude

The transformative and replenishing powers of healing presence and gratitude await us on the other side of breakthrough. They display the grace and resource breakthrough can provide.

Let's talk first about what healing presence is and why it's important. How is it that having another person acknowledge our pain and accompany us through that pain—whether we are patient or clinician—mobilizes within our bodies, our psyches, our minds, capacities that we are not necessarily able to mobilize alone? What is the demeanor, the stance that can be held by another that allows the person who is ill to mobilize something new? That would be healing presence (Frank 2002).

What I'm thinking about is the ability to stand in front of a person who is suffering and to remain centered, to remain nonreactive, to remain clear. Being in the presence of suffering stirs reactions of all kinds in us. There may, for instance, be repulsion—as from the smell of a burn unit, or severe disfigurement from accident or disease. There may be simply the painfulness of erratic or bizarre behavior of another that we're not sure what to do with, or that stirs painful memories. There may be reactivity—annoyance, resentment, anger.

The critical skill at this point is being able to allow one's reactions to happen, without letting them distort our attention or break our connection with the person in front of us. A more subtle level of this is not being judgmental. Judgment is a form of distancing. It's a labeling process, one intention of which is to contain or bound

Into the Field of Suffering. David Schenck with Scott Neely, Oxford University Press.
© Oxford University Press 2023. DOI: 10.1093/oso/9780197666739.003.0006

something we find troubling, and another of which is to save time. But if we dwell exclusively or even primarily in our judgments of good versus bad, right versus wrong—even true versus false—we will often be blinded to what's unfolding in front of us. Everything that's happening is true to human being. Holding a healing presence means not being stuck in judgments of a moral, or clinical, or scientific kind.

That's not to say that we throw away our analytical mind. Our analytical mind is something that needs to stay present. But it needs to be an open and flexible mind. Healing presence involves being still, calm, nonreactive, and nonjudgmental. Sometimes the word "accepting" is used, but it's unfortunately open to misunderstanding. It does not mean that whatever anyone does or wants to do should be allowed. "Accepting," properly understood, means, "I accept that this is the reality. I accept that this person is broken in these ways. I accept that this behavior is exactly harmful in these ways. I accept that this person's behavior may be driven by all kinds of factors that I don't understand." I am talking about the willingness to look and to see—and to stay put after we have seen what we've seen.

An important distinction can be made: There is a difference between the *call* to relieve suffering and the *need* to relieve suffering. This points to a central dynamic of the growth of a healer throughout the course of their career. The more self-contained and centered a person is, the less needy they are in the given moment, the more likely their presence is to be healing. All the work we do with our vulnerability is aimed at strengthening the presence we can offer another.

Many people in healthcare who are called to do the work to relieve suffering also have a deep need to relieve suffering. Indeed, depletion and burnout for many is the feeling that the world is refusing their offers of healing. But what in fact is being refused, at times appropriately, is their *need* to heal. A willingness to serve healing is something else entirely. The remarkable craniosacral therapist I mentioned in Chapter 2 once said to me in an interview, "It's incredibly presumptuous to go into the room with the intention to heal somebody." Given the truth in that, what is the proper approach? I remember Brugh Joy talking about healing presence this way: "You put your hand up and you bring your intention to heal to the surface of your skin. But you don't project that intention out toward the other person, or into the other person" (Joy 2008). In other words, the intention to serve and to heal is present and, if the other person is receptive to it and is ready for that connection, the healing that is possible will happen. (Cf. Remen 1999.)

I. "Our Brain Doesn't Even Have the Words . . ."

To illustrate what I mean by "healing presence," and to prepare for our discussion of gratitude, I want to recount an interview I did when doing the fieldwork for the book *What Patients Teach* (Churchill, Fanning, and Schenck 2013). My colleagues Larry Churchill and Joe Fanning and I interviewed 58 patients, many of whom were patients of clinicians that Larry and I had already interviewed for our previous book, *Healers* (Schenck and Churchill 2011). But the cohort also included 18 patients who were receiving hospice care.

The intention of the patient interviews was to have patients talk about how they understood healing relationships with their clinicians. Specifically, we were interested in what it was about their clinicians and the people who worked with them that contributed over the years to healing experiences, and in what clinicians had done that was an impediment to healing. With the hospice patients, who were facing a terminal diagnosis and had a prognosis of six months or less, the temporal focus tended to be more in present time, but the range of topics was often broader. The hospice patient whose interview I'm going to recount was in her mid-50s and dying of abdominal cancer. She had clearly been an extraordinarily capable corporate executive. She was intelligent, vigorous, insightful—at one moment discouraged, in the next hopeful.

This interview was an altogether remarkable experience for me and for her. It occurred in the patient's home and lasted just short of two hours. Along the way, we talked over what she saw and felt and knew about the changes in her life since her terminal diagnosis, including the nitty-gritty of living while dying. We talked about the many kinds and levels of healing she had experienced. The events in her story are not uncommon; but her insights about these common experiences are sharp, trenchant, and embody wisdom about things seldom discussed.

I bring this conversation forward for what it has to reveal about how presence alone can be healing—in large part by helping people mobilize their own capacity to heal themselves. But it also picks back up on the stories I told in Chapter 3 about my father's death, and about Mr. Jones, and my learning how to turn those experiences into a resource for people who were dying. In all my ongoing work with hospice patients, across three decades, I was steadily aware that these patients were giving me opportunities to heal my own wounds. And certainly in this case, as I was offering this woman something that I believe was healing for her, I was, at the same time, receiving remarkable gifts.

I had already done two long hospice interviews by the time I got to this patient's apartment in the early afternoon. I was tired, and badly in need of a bathroom. Not at my best, in short. I knocked on the screen door. It was closed, but the main door to the apartment was open and there was this voice from down the hall. I couldn't see anyone, but I could hear barely someone saying, "Come in, I'm down here." I walked in and went on to the den. The patient, the interviewee, was lying sideways on a small couch. Ms. Brown had a big pillow behind her, resting on the left arm of the couch; a pillow between her body and the back of the couch; a couple pillows under her knees; and her feet were hanging over the couch's right arm. She looked to be in considerable pain, judging from how she held herself and how she moved. Her first words confirmed this.

"This is a bad day. I'm in a lot of pain. I thought about calling to cancel this. But, well, we'll try it; we'll see if we can do this." I said, "I understand completely. We can stop at any point. It's completely fine." I added, "Before we start, if you don't mind, could I use your bathroom for a moment? I've been interviewing all morning." She said yes, I went down the hall as indicated to the bathroom.

I come back into the room, sit down, and immediately notice that the patient has become a good deal less lethargic. She looks at me very directly and says in a sharp voice, "I'm not comfortable you went in my bathroom." I say, "Oh, I'm very sorry." She says: "All of my narcotics are in my bathroom. I can't go upstairs anymore, and I keep all my narcotics in the bathroom. I don't know you from Adam's house cat and you've been in there with all my drugs." I'm of course thinking to myself: "What a horrible way to begin this interview!" I'm also wondering what job this person had had, for she clearly has interpersonal combat skills, which are still in play and easily activated.

But she had a good point—she knew nothing about me. She knew I was recommended by the hospice agency and that I was writing a book. But, truthfully, from her standpoint, that didn't mean squat. And I know this. I say,

> Whenever I go visit someone who is a hospice patient, the first thing I do is go in the bathroom and wash my hands. It's how I was trained 25 years ago. I always do it; it's a matter of showing respect for the patient. Also, as I mentioned, I've been driving around all day, and I really did need to use the bathroom. I'm terribly sorry this made you so uncomfortable. If you would still like to go ahead with this, we can do that. But if you are uncomfortable with me and prefer that I leave, we can do that as well. It's up to you. What would be best for you, right this minute?

Ms. Brown explained again, tartly, why this had made her uncomfortable, and I again acknowledged the point about the medications. Once more, I gave her the choice about what to do next, and she said somewhat begrudgingly, "Well, all right. I don't like it. But we'll go ahead." We began to talk—awkwardly at first, to be sure. I asked her the question I always ask right away in a visit with a hospice patient: "How is your day today?" I don't say, "How are you?" because the easy answer could rightfully be: "I'm dying, you idiot, how do you think I am?" I always specify *today*. I'm not asking for any grand pronouncement of anything—of prognosis, or of the meaning of life and death. And I ask about them, not their clinicians or their family—which are often loaded subjects for the dying. It's just: "What's your day like today?"

This invites a person to say what this woman said: "I'm in a lot of pain; I'm uncomfortable." It gives me a chance to watch the patient speak, to see how comfortable they are, how uncomfortable they are, how what I'm seeing matches with what they're saying. It gives me an instant index of credibility. And it typically surprises people, because the experience of the dying is that most people who come to see them don't want to know how they are, and will not generally ask—that direct question being too scary. And if people do ask, they generally want a reassuring answer. Either that, or it's a medical professional, who is asking how you are with a special interest in your symptomatology.

My experience has also consistently been that it is almost always a question that opens a good space on both sides of the conversation. And thank goodness, that was the case here, too. Next, I explained more about our project and asked her to tell me how she had gotten started with this hospice. I gave her our interview question sheet; she looked it over, and on we went.

We got immediately into a litany of the terrible things that had happened to her in the healthcare system. Her primary complaint was that no one would give her an honest prognosis of how long she might live. Everyone, including her oncologist, evaded that question. And on we went from there. When she would go in to have tests done, she unsurprisingly found the machinery terribly uncomfortable. The staff were often laughing and joking. There were inappropriate comments made to the patient, as well as back-and-forth among staff. She felt she had no shred of dignity left by the time these tests were done.

Then Ms. Brown went back to her primary oncologist, and we got into the complexities of that relationship: "She didn't know what to do with me, because I always wanted to know the real truth." Or: "I kept asking for more information, and this was not what she usually got from her patients." She did have awareness that she might not have been the easiest patient. I found myself sympathizing both with her and with her doctor.

I was validating all this as we went along. She seemed a reliable witness. I trusted her account right away. And I was quite familiar with the special problems our healthcare system presents for dying patients and their clinicians. As the conversation moves along, the patient discards a couple of her pillows. And, because she's so vested in this story she's telling me, she's turned to face me. When we began the interview, I was sitting in a chair opposite the couch, and she was looking at the wall opposite her, her sight line perpendicular to mine. She wasn't even looking at me. Though she's clearly still in pain, as her narration continues, she sets aside more pillows and faces me even more directly.

A decisive turn in the conversation came when I said, "Tell me about the other people who've been important resources for you, besides the folks in hospice." She began then to talk about a few of her friends. She was single and lived alone. These individuals were her essential companions throughout.

Ms. Brown described friends who took her to her doctor's appointments, who took her to all her treatments, who helped her with her finances, and who were with her when she was dealing with the worst side effects. As she talked about these friends, she discarded another pillow and was sitting with her feet flat on the floor leaning forward. We went on to talk about her children, and she continued to get more and more animated, and was moving with minimal apparent discomfort. She was sitting up so alertly and so engaged in the conversation that if you hadn't known she was sick and you had walked into the room, your first thought would certainly not have been, "This woman is in terrible pain and will be dying soon." You would've thought, "What a lively conversation these two are having!"

Next, she talked about her grandchildren. And at this point, her tone and posture shifted once more. She was leaning forward over the coffee table that was in front of the couch, because she wanted me to understand exactly what she had to say about these two girls.

Before going into detail about the remarkable vignette the patient related, I want to make a few points about healing presence. This interview had certainly started out inauspiciously! I had begun by inadvertently making the patient decidedly uncomfortable. And her immediate response to that was aggressive. Not inappropriate, but a bit over the top. We went from there into all the problems she had had with her healthcare. We talked a lot about her physical pain. During all this conversation, she was markedly uncomfortable.

There is nothing that I offered her, could offer her, during that time, other than: not leaving, not being reactive; recognizing the truth in all she was saying; and inviting further conversation. If you had filmed this with no sound, you would have filmed a person sitting on the couch packed with pillows in great pain, who then shifted around, placed her feet on the floor, leaned forward, and engaged in animated, vigorous conversation. You would have been filming healing—of a certain kind for a certain period—and healing presence. This was for me clear, graphic, physical validation of the power of being present, staying centered, and not being reactive; of encouraging, recognizing, listening, and affirming. In the process, this patient began to tell a story that was more complex about her physician; she began to express enormous gratitude for her friends and appreciation for her friends; and she brought forward this remarkably intimate and intricate story about her grandchildren—all of which was healing and animating for her, and enormously meaningful for me as well.

"My granddaughters were visiting," Ms. Brown says, "and, can you believe it? They're 10 and 12 years old, and they didn't understand that oak trees grow from acorns." I say: "No, I don't believe it, I'm with you." She says, "You know, I told them that I had something that I needed to talk to them about. 'Nana has something she needs to tell you; come sit out in the garden with me.'"

She says to the two girls, "Look out there, you see all those acorns?" "Yeah," they say. And then she says to me,

> I talked to them about acorns and about how the acorns had fallen from this big tree right here. And I told them, "This tree, let's talk about this tree, it started with this acorn, and it grew and grew, got bigger and bigger. And then came the day it opened up to become what you see right now. And now it pours all these acorns out on the ground." And I asked my granddaughters, I said, "You know what will happen to this tree as it gets older? What will eventually happen to this tree?"

The girls say, "Well, Nana, the tree will die. We've seen trees die." The patient says to them,

> You know, your Nana is like this tree. I started out like it did as an acorn, and I grew and grew, and eventually, I got to be like this tree right here, big and strong. Many, many wonderful things happened in my life. I did all these things. I poured acorns all out into the world. Well, girls, you know that your Nana is sick, right?

They say, "Yes." The patient says,

> The sickness that your Nana has means I'm going to die. I'm probably going to die in the next few months. I'm going to be like this

> tree, I've grown this big—and then, like this tree is going to die, I'm going to die.

One of the children asks her, "Are you afraid, Nana?" "No, no I'm not afraid of dying." They had made a simple and quick transition, as children will: "Okay, what are we doing next?" Without any discomfort whatsoever.

After she gave this detailed account, this woman and I had a wonderful conversation about children and death. I said, "You know, we spend a lot of time in our culture protecting children from death. But in many ways, children are far more comfortable with dying than adults are." I asked, "Is this your experience?" And she said this very fascinating thing: "I feel like children are less uncomfortable with death, because they are closer to the place that I am going than the adults that I talk to. *They've just gotten here*." It was terribly powerful.

At this juncture I quoted (slightly inaccurately) Wordsworth's "Ode: Intimations of Immortality," making the point that children "come into the world trailing clouds of glory." Then, she said another amazing thing:

> What I have found is that there are not enough people who recognize that we all want the same things. You know, we really do all want the same things out of life. *And what we need to do is to help each other get them.*

A vision of kinship and communion and interconnection as she approached death—very beautiful.

Ms. Brown went on:

> There are lots of things I still don't have any answers to. And that was another thing the girls and I had a chance to talk about—about how what happens to us after we're gone is probably so amazing that *our brain doesn't even have the words to describe it.*

This courageous woman had gone from being folded in on herself and her pain, folded in by her pain, folded into her body, into this full and generous sharing of the most profound pieces of her life. For me, who's been a student of death for decades, this was marvelous beyond words. By then, we had been talking nonstop for nearly two hours. This person who wasn't sure that she was even going to be able to do the interview had steadily gained energy—had brought forth more and more complex stories, deeper and deeper insights.

Realizing Ms. Brown was likely far more exhausted than she knew, and worn out myself, I said, "Our time is done." I stand up and, to my surprise, she stands up as well. We begin to walk down the hall toward the front door. She says, "Come in here for a minute." We go into what clearly used to be her dining room. There's this huge table covered with papers and files. "This is the war room," she says, "where I fight the insurance companies." I'm picturing this executive in battle, equipped with printers, computers, and all this paper. And I'm noting to myself that it's insurance companies who are the war enemy, not the disease that's killing her.

We walk together to the front door. We've had this totally remarkable conversation and shared a startling intimacy. We turn and look at each other. At this moment in a normal conversation one might say, "I'll see you next time." Or, "I'll let you know what happens with this interview." But neither one of us says anything; we just look at each other. And I can tell that she's thinking the same thing I'm thinking, which is that she'll be dead in less than a month. We know there's no further conversation; we know that this conversation is the only one we will ever have. We know that this is not a moment that is for anything other than itself. We know that we have experienced something that, in and of itself, for both of us in our different ways, is pure gift and pure healing.

The especially amazing thing is that *we both knew this*. We both recognized it, but didn't even try to name it, didn't mess it up. We said thank you and goodbye. A remarkable experience.

Sometimes one has the experience of being with a person who's suffering or dying and becoming aware of a huge range of responses happening inside, all at once: the compassion, the potential for healing, the grief, the joy, the pain. But what was especially remarkable about this encounter was that the dying person herself was holding this entire range as well, at the very same time that I was. Her own capacity, and here I am projecting, had been—we see this often—transformed by her illness. What she could hold had been expanded greatly by her very suffering. My guess is that the hell-bent executive of three years before might not have been able to articulate her gratitude for loving friends, or tell that wonderful story about her granddaughters, in quite the same way.

I've often had the experience sitting with someone who is close to death and watching them run through the whole range of human responsiveness. But I had never been with someone who was experiencing that full range, *and was aware of it*, while I was holding the full range myself, and aware of it. The conversation with Ms. Brown was a shared, direct experience of human finitude and limitation, as well as of generosity and healing, illuminated by her impending death and my own eventual demise.

I should also say that this to me is a paradigmatic situation for realizing how much gratitude we can have and should have for those people that we serve, for those people that we offer to help when we respond to this vocation. We're going to talk in detail about gratitude in a moment. But I want to underline the fact that this woman gave me, during that conversation, in her willingness to stay in the conversation, and in what she offered in reflecting on her dying process, things I could have gotten nowhere else—and could not have discovered on my own. All that, along with forcing me to face my own reactivity: "How dare you ask me why I went in your bathroom?!"

The profundity of the trust that she ended up offering by the time she told me the story about her granddaughters came out of mutual recognition. I offered her recognition of the realities of her dying;

and she stretched past her own reactivity to recognize a capacity and a need in me as well. And she followed that with the gift of her own experience, the gift of a brief portion of her life. We do well in the process of living the healing vocation to continue to be on the lookout for opportunities to both feel and express our gratitude to and for the people we're serving. They are doing things for us that we could not begin to do for ourselves. They are guiding us through ranges of the human realm we wouldn't otherwise know. And doing this with remarkable courage and generosity.

II. "Where There Is Gratitude, There Is Healing"

As I noted in Chapter 1, one of my mantras is: "Where there is gratitude, there is healing." Gratitude is one of these overlooked human capacities that can seem insignificant, but which is infinitely worth focused attention and experimentation. It is, for instance, virtually impossible to be feeling negativity and gratitude at the same time. Try it and see—most remarkable.

Gratitude is something that is also almost always available. Sometimes, we must admit, it's hard to feel loving or compassionate in challenging situations, or even in everyday life. Sometimes it's hard to muster even minimal empathy. But, almost no matter what, there's something going on that we can be grateful for—even if it's contained in the uncharitable thought that it's the patient who's in the bed sick, and not me. Or even: "Isn't the body astonishing, even as it goes through this cancer process—isn't this an amazing and awful thing?" Or: "It's good to breathe; it's good to be here; it's good to have a body." Or: "Geez, the clouds are interesting today." Whatever can give us an inkling of an invitation to gratitude will help us pull ourselves out of reactivity and negativity.

Gratitude is oddly parallel to belly breathing in its relation to reactivity and anxiety. If you do belly breathing properly and come deeply into your abdomen, you can defuse the anxiety reaction in the body. Simple belly breathing is standard training for people who have anxiety attacks because they're incompatible processes. And control of your breathing is something that, as I've said before, is almost always available to you.

The parallels with gratitude are (1) it is always available and (2) it is incompatible with, and has the potential to cancel out, negativity of all kinds. Gratitude taps generosity. Gratitude is a path into the heart center that is, in most ways, less arduous than compassion. Gratitude in its more rudimentary forms is more accessible because it's admittedly self-oriented. I'm grateful for "X" because some good thing is happening to me. But gratitude's special power is that, at the very moment I'm happy for myself, I'm also thankful—I'm acknowledging that I am receiving.

And then, subtly, almost unnoticeably, gratitude can come into other dimensions. Here can move a profound realization of interconnectedness, bringing with it a sense of having been gifted or graced. Further on, there is the growing realization that one has done nothing to earn or deserve the truly greatest gifts that have come one's way.

Gratitude has a universality to it; it's helpful in virtually all times and places and circumstances. It's also something that I find people in our culture are happily surprised by, when it's directly expressed. Not the perfunctory, "Oh, thank you for that." I mean saying things like: "I'm grateful for what you did just then." Or: "I'm grateful for what's happened today." My experience is that such phrases almost always alter the nature of a conversation. It's as though people feel invited into a particular type of intimacy—not a threatening intimacy, but an opening, an invitation to share in the generosity of the world.

A wonderful resource is a short video narrated by Brother David Steindl-Rast, the title of which is "A Grateful Day" (Steindl-Rast 2017). It's just over five minutes long, and it goes lovingly through a myriad of typically unnoticed things to be grateful for: running water, electricity, flowers, smiles, the sky. He closes by saying, "Let the gratefulness overflow into blessing all around you. And then it will really be a good day." It's a photo montage—the images flash by of people from all over the world. Many, many different cultures, skin tones, expressions, ages, activities. And the beauty of the way he teaches this! A delightful set of reminders of how utterly ordinary gratitude can be, how miraculous what we tend to think ordinary is, and how it is gratitude that gives us access to that miraculousness.

There's an exercise to consider here as well. The loving-kindness meditation, or *metta* meditation, mentioned in Chapter 1 has a structure for offering wishes for well-being to others that can be utilized to guide a systematic "gratitude practice." Alongside Steindl-Rast's gratitude for food for the day, gratitude for water, air to breathe, and the earth to live on, there can be gratitude for the people with whom we work, gratitude for our patients. The intention of this meditation would be, as it is with loving-kindness meditation, to exercise and invigorate the heart center. We can call to mind systematically the people we are grateful to have in our lives. We could think of mentors and models who showed us the way. Colleagues and friends who support us. Families and partners and other people we love. And, as we advance in the practice, we can at times become grateful for people who annoy and even hurt us, treating them as teachers along the way.

A gratitude process or gratitude practice is in addition a further way to acknowledge our interconnectedness, to pick it up intentionally and strengthen it. Interconnectedness is at once the source of our suffering, and the source of our capacity to offer healing and compassion. In gratitude, we acknowledge that we don't exist without all the rest of the universe coming together to produce this moment where we find ourselves. In speaking gratitude, we are at the most fundamental level acknowledging the fact that the universe that produces us is a gift to all.

I've already talked about having gratitude for the people that we serve, for the patients, for the clients, for the people who find themselves in situations of great vulnerability who allow us to come into their presence. But in healthcare we also have the opportunity to express gratitude to other people who share our vocation, to the people who are willing to be with us as we are with people who are suffering. We have the opportunity to be grateful to our team members, to our colleagues, because none of this is solo work.

We can push this further. We can try to stay aware that, like ourselves, the people who join us in this work, at the bedside let's say, bring their own woundedness with them. They bring their own complexities of vulnerability and capacity; otherwise, they wouldn't be putting themselves in these kinds of situations. If we are stressed and taxed by the suffering and brokenness that we're facing, we can be sure they are as well.

Our heart can go out to them because we know the challenges they are facing; we feel those same challenges in our own bodies. This is an ongoing opportunity for gratitude: For the people we work with who manage to remain generous and compassionate and energetic as they go about our shared work.

***

I opened this book with an invitation to you to explore with me the healing vocation, this call to turn the field of suffering into a field of healing. As a person who has been called to caregiving, you have stepped forward with willingness. The very fact that you have engaged this book suggests that you are engaged in deepening your own sense of vocation. I want once again to express my appreciation to you for this. As Cheri Huber taught me long, long ago, "Willingness is the key." Without willingness, nothing happens. Not all are willing. Willingness and generosity—these are the keys (Huber 1999).

You have stepped forward and are continuing to step forward. I invite you to continue the conversation with your colleagues, with your patients, with your clients, with people you encounter who are in situations of suffering. The work is demanding; it necessarily depletes us; we may even burn out. But we can break through; and we can help one another do that. I invite you to continue the process of exploration through the exercises that you find here, and to continue to explore the mysteries of healing presence.

These mysteries challenge us. These mysteries call us into the depths of compassion, where we find that there is a power in us we do not understand, whose source we do not grasp; but a power which, when we are in the presence of people who suffer, is called forward in us. It is a power which heals us and those we touch.

The tasks that must be done to promote healing are often grim ones. But participating in healing can be a work of deep and abiding joy. An odd word in this context in some ways, but the very possibility of healing is itself totally remarkable. The gifts that we are offered by the people who come to us are remarkable. The gifts offered to us by the people with whom we work, well beyond the bounds of expectation, are a source of deep healing and joy. I thank you for your willingness to be on this path, and for the courage that you have shown in coming this far—and I look forward together with you in hope.

PART II
THE DIALOGUES

Deepening Capacity

The following dialogues invite you to join in the continuous exploration of what healing is and how we can sustain one another as healers. These conversations were held between the author, David Schenck, and his collaborator, Scott Neely. Neely is a pastor and community advocate.

Dialogue allows us to practice the skills of attentive listening and apposite sharing. In dialogue, partners accompany one another into ever-evolving, ever-deepening understanding. We close the book with these five short pieces to remind us all that the conversation continues, that learning never ends, and that we are not alone.

Dialogue 1
Honing the Exercises

Being in the Midst of It

Scott Neely: You talk about exercises of awareness as personal practice for addressing physical or emotional pain in one's body. And you indicate that one can easily make the move to doing variations of these practices on the spur of the moment, during a conflict or in a stressful setting. Could you talk more about how these awareness practices can be successfully transferred?

David Schenck: The part that can most easily transfer is shifting your posture—straightening your spine and planting your feet. Simply doing that, even if you've never done any of these practices, tends to have a calming effect. Why is this? If you talk to a chiropractor about the spine and alignment of the spine, you get one explanation. If you talk to Ayurvedic medicine people about chakras and various yoga meditation techniques and postures, you get another explanation. If you talk to Taoists and Qigong masters about the central channel and the alignment of chi, you get still another one. Ditto for neurologists trained in allopathic medicine. Zen Buddhists will say, "When you sit, you are Buddha." That's their explanation. What's clear across cultures and explanations, across the history of thousands of years of practice and healing, is that aligning the spine, calming the breath, and planting your feet—or if you're in lotus posture, settling your hips into the ground—is a soothing and centering thing for human bodies and human beings. Almost anywhere you are, you will be able to sit up straighter, or stand up straighter. But the second thing about this is that it gives you

Into the Field of Suffering. David Schenck with Scott Neely, Oxford University Press.
 DOI: 10.1093/oso/9780197666739.003.0007

something active to do with your body while wild stuff is going down around you. You're not in a physical posture of receiving what's going on, not at the mercy of whatever's swirling. You're making an active movement with your body. You are, quite literally, taking a stance.

That doesn't mean you're going into Jedi defense mode. It only means that the process of being aware that you can move, and making that move, changes your emotional experience. Movement demonstrates to you, in the most convincing way—in a bodily and concrete way—that you have control over at least two things: your body posture and your breathing. And this realization opens you to a more profound process of taking up the negativity or the reactivity that is rising in you as something to learn from—taking it up as data, as information you can use.

But for it to become truly effective . . . that's where having an established practice and having worked on awareness exercises ahead of time comes into play. Then you have developed capacity to call on. But even if you haven't done any of that preparation, you will find that being able to change the posture is a powerful thing. People will say, "I don't have time for that." To which I reply, "How much time does it take you to straighten your spine and plant your feet?" It takes no time. Two seconds, one second. You can do this in the middle of a resuscitation, a code. It can be done anywhere.

Scott: I hear you saying that when you utilize these practices, you're able to be in the meeting, the code, whatever it is, with a different capacity. To say that I don't have time to do it is to say, "Please don't add any time-consuming activities to my routines." But expanding capacity is a movement to expand the moment itself.

David: Right. Which in turn makes room for more activity. *It gives you more time, not less.* What I'm talking about is a *way* of doing something; it's not an additional task. It's changing the way you do what you do.

And the more you've practiced this, in and out of work settings, the more your body will become habituated to it, with the result that you'll be able to respond more effectively when a crisis arises. If you've done this posture many times at home and on retreats, as soon as you straighten your spine and put your feet on the floor, your body is saying, "Oh, right, mellow. Yeah. All this wild stuff going on, but the message here is mellow."

Scott: You went exactly where I was thinking of going. This is not something that you have to have already done to benefit from it. But the more work you've done with it, the more benefit you may gain. Which is logical. The more you practice sticking someone to insert an IV, the better job you'll do next time around.

David: Exactly.

Scott: We know when we go into challenging situations, which we're choosing to do, that intense emotion is going to be moving. The more awareness practice we've done focused on emotion and pain, the more ready we will be for whatever emotions are running. Suppose, for example, you're going down the street and somebody grabs your bag and starts to run. You can always yell, "Stop thief!" But if you happen to have been training as a world-class sprinter for the last few years, you can yell and run and maybe outrun the person, right? The training is of great benefit, even if there's a response that's available to all of us.

David: The thing I would emphasize is that if you're going into an intense situation, you know there's going to be an internal response, you know there's going to be a *bodily* response. Depletion and burnout can here be such powerful teachers. Depletion and burnout are manifestly physical as well as psychological processes, though we still manage to be surprised by this. In healthcare, we shouldn't be surprised. We are working with bodies that are suffering, and that's inevitably going to be absorbed into our bodies. Our emotional reactions will move through our bodies as well. Further, they will manifest themselves in our bodies. Being present to suffering is an embodied process.

How we train our bodies and minds to relate to our emotional reactions is thus incredibly important. It's not nearly enough just to say, "Suck it up." Which is about all the advice most medical providers in most disciplines seem to get, in their training and afterward. And it's not enough to explore only the psychological dimension. There are critical physiological dynamics, as well. In my opinion, profound work with depletion and burnout in relation to moral distress must include bodywork, if it's going to be transformative.

SCOTT: It's funny to think about the medical domain, which is dominated by technology and intellectual constructs, being at bottom about the caregiver's endurance. Providing healthcare is a very physical experience, yet we don't pay much attention to this most basic fact.

DAVID: Think about the way that healthcare professionals are trained to think about bodies. They're trained to think about them in scientific and objective terms, and as physical systems. They're not trained to think about the body as something that we indwell or inhabit—that, in fact, *we are*. So often the reason patient care goes askew is that, dumb as this sounds, *we forget that the patient's body is the patient*. We dissociate the "technical body," which comes to us through our machines and measurements and lab values, from the "living body" that is the actual patient. It's not altogether surprising, then, that healthcare professionals would also internalize that split between the lived body and the body understood scientifically, and that this in turn would mask processes in their own body and obscure their own lived bodily experience.

Approaching Physical Space

Scott: Here's a silly image but go with it for a moment. You're in this medical center and there are all these ways of learning and engaging and trying to help your patients. And yet if you push on this secret door, it's as if you can change the whole space that you're working in, change all its dimensions. And you can do this by paying attention to angles and stances and spaces, which are right in front of you, but which rarely had sustained awareness brought to bear on them.

David: Here's an application of your thought experiment. I'm discussing with a nurse a difficult interaction she's had with an attending physician. And I say to this nurse, "Where did this conversation take place?" "By the nursing desk." And I say, "Did it take place in that narrow space, where that shelf is jutting out?" "Yes." I say, "So you weren't able to keep a suitable distance away from this person? You were backed into that little corner over there, weren't you?" Answer to that is yes. "So next time you have to talk to Doctor 'X,' don't stand there. How about over there by the pillar? That's a space where there's breathing room." "Well, usually I'm talking to him after morning report, which happens right there." And I say, "I hear that. But let's think about this. Will it be hard to say, 'Dr. Smith, can we step over there? I can't hear you. Can we move away from here?'" Which simple movement makes the actual, available physical space a safer space—emotionally and physically—to have a conversation in. And we're hoping for a conversation that won't remain stuck in the throat or the chest after it's over!

Or take another conversation I had with an attending physician about a difficult conversation with a family. "Which room were you in?" "Oh, it's that room that's always crowded." Or: "Oh, it's that room where there's a dark corner and a chair in the back. And the dad was back there and didn't want to come out." Or a conversation with a resident about a confrontation in an ICU room: "It was across the patient's bed, and it was like we were doing a tug of war with the patient."

Because we are bodies in a room, in a space, the rapport of those bodies and how they interact is an important dimension of our emotional experience. One reason to think carefully about the geography of the room is to get a better understanding of how and where the emotions are going to flow once things get intense.

Scott: What I'm hearing is that when we say, "Could you come over here and talk?" or "Let's step out of this room for a moment," we are reclaiming not just our autonomy, but also the very space we need to do our work.

The message of these simple yet powerful moves is, "I belong here. I have something important to contribute."

David: Another example: My hearing is bad on the right side, as you know—and this is always a great excuse. "I've got a bad ear. Could you step over here, please?" And virtually no one is going to say, "Tough beans, buddy. We're staying right here, even though you can't hear." There are a few people who might do that. But it's a small percentage and, if nothing else, what you learn from that is, "Okay, this is one of *those* guys." And for them you deploy a whole different repertoire of skills and approaches.

Scott: This is a bigger point than I grasped at the beginning of this conversation. The fear is, "I don't want to cut people off or step away; I'm here to help." But the move you're suggesting does not cut people off. It doesn't end a conversation. You are continuing the conversation, but on new terms: Step into a different space. Sit up straight. Claim the space as your own.

Dialogue 2

The Exercises as Spiritual Disciplines

Scott Neely: Why call these exercises spiritual disciplines? What's the value in that framing?

David Schenck: The initial inclination to refer to them as "spiritual disciplines" came from the fact that they draw heavily on skills and practices developed in Eastern and Western religious traditions—meditation practices, contemplative prayer practices, chanting practices, liturgical practices. That's one element of it.

Another has to do with a broader sense of the term "spiritual" we've been elaborating since the first chapter. We have used the word "spiritual" before to indicate multiple levels of connectedness. And, speaking on the level of the person, we've talked not only about the psychological or social or physical dimensions of life, but the spiritual dimension as well. Those deep relationships with the environment, with the universe, with the sacred so central to our lives.

This leads into the next thought: The idea that these "disciplines" have the potential to alter the total quality, nature, and richness of our world. These exercises operate on our habitual ways of being, which impacts literally everything we experience and everything we do. "Spiritual" is intended to indicate that global potential for transformation.

Scott: What I hear in the use of "spiritual discipline" is a reclaiming of a catalog of traditions and centuries of human activity for the sake of doing one's work well, rather than being confined by a dogmatic or doctrinaire framework.

Into the Field of Suffering. David Schenck with Scott Neely, Oxford University Press.
 DOI: 10.1093/oso/9780197666739.003.0008

But beyond that, if we choose words like "cosmic," certainly that has the sense of something that is more than global and interconnected. If we choose a word like "wholeness," and religious corollaries to that, like *salaam* or *shalom*, we are tapping into spiritual values expressed in various traditions. These are words that mean exactly what I think you're talking about, which is the possibility of nurturing and generating goodness through the web of connections that is our life. That's what I hear you describing when you talk about healing as a movement toward greater wholeness.

David: That's great. You've said it very well. We're situated in our lives not simply as psychological beings. We're in a cosmos. We're intertwined with a vast range of environmental and social factors, and the immensities of biological life. By using the word "spiritual" as a general term, we intend to point to the fact that we situate ourselves in the universe that is hugely complex and well beyond our understanding. We're issuing an invitation to look at things as comprehensively as possible.

Dialogue 3
The Essential Skill of Advocacy

Scott Neely: When I hear you talk about your colleagues, the people you work with, I hear great care, and a desire to affirm and support. I hear concern that so many are in unsustainable jobs and depleting work settings. And in that desire and that concern, I also hear a call to advocacy.

David Schenck: When I first started debriefing nurses, I kept going back to this one unit time after time. And soon enough, I began to get disconcerting feedback: "Oh great, ethics is back. But why even talk to this guy? Nothing ever changes." And of course I'm wondering what this is about. I finally understood the message to be: This is a system problem, an institutional problem, and this guy isn't going to do anything about that. All he's doing is listening and trying to make us feel better; but it doesn't change anything about our job, day to day. And these same exact bad things are going to happen again next week, if not tomorrow.

And they were right. The exact same things *are* going to happen again, and probably soon. So now when I go and talk to these groups, I say, "There's two sets of things that we need to do. One, I can talk to you about advanced skills for dealing with depletion and moral anguish. I can teach you those things; we can work on that. You can develop and expand your capacity. But two, the other thing we need to do is talk about advocacy. Because the sense of powerlessness and being trapped and having to passively endure situations, is unbearable over the long haul. Career death. And 'death,' death."

We need to get people to go further in their thinking about systems and institutions. To think of institutions—hospitals,

Into the Field of Suffering. David Schenck with Scott Neely, Oxford University Press.
 DOI: 10.1093/oso/9780197666739.003.0009

clinics, nursing homes—as systems they might conceivably have an impact on. "Let's talk about the things we need to discuss with your attendings." "Let's talk about the things that we need to discuss with your nurse manager." "If you're afraid to talk to your nurse manager, I will go talk to your nurse manager. If your nurse managers are afraid to talk to your attendings, we'll figure out a way to get them together." The goal being to have people, who with reason assume they have absolutely no voice in the system, come to see that they can join the conversation. Or, if necessary, demand to be included!

I've been observing changes in healthcare over the last decade, in and out of hospitals. Shifts in economic drivers, changes in the third-party payer systems, the impact of population health measures on the daily life of nurses and physicians. Such enormous changes—and few of them to the good for patients and patient care. Our nurses and doctors are being driven into very difficult and often dangerous situations. Just look at the failures in the PPE supply during the COVID-19 epidemic. So obvious; so inexcusable. At some point, it's all going to come apart.

Scott: Not to neglect the physicians, who I know are impacted by all of this. But it's striking to me that you talk in such detail about nurses. They are the group that comes to my mind as well. I know many who work in various healthcare settings, and all are under tremendous amounts of pressure, working under administrative systems forever trying to get more out of them, without overtaxing the budget.

What do I see in this? The healthcare system brings them into a relationship with patients, who are asking for healing because of crises in their own lives. That seems on the face of it to be an unqualified good. People coming into relationship to help each other. The shadow side is the economic players who see multiple profit-making opportunities there, and that create and control those structures in which you must work. Schooling, licensing, facilities. No other options. You can't just set up your own shop.

This means that healthcare becomes an extractive industry. People's labor is mined, to the point of having their own health endangered. Yet the call, the vocation, pulls people into the work deeper and deeper. They fight against the rising tide and keep doing the healing work, to the joy of the administrative system, as their workplaces become more and more extractive. And lest we lose sight of the still broader context, that's hardly the only driving system. There is big pharma, the tech companies, and insurance conglomerates pushing it all. These are massively depleting forces that the healing work collides with. How could that not lead to widespread burnout?

The question I want to ask is: Were we to posit advocacy as an exercise, how might we present that?

David: Let's start here. We know that a critical component in moral anguish, depletion, and burnout is the sense of being trapped, of being powerless. We know that there are always institutional vectors rendering us powerless and trapping us in morally distressing situations. Vectors that make already challenging jobs even more difficult. And this is just where the impetus for advocacy comes from. Advocacy, as I use it here, means finding ways to challenge the people and alter the structures that undermine, or make impossible, the healing work you are called to do.

Advocacy is essentially a set of community-organizing challenges: cultivating relationships, making connections, building networks. It cannot primarily be a matter of individual effort; but it does begin with individual initiative. Oddly enough, though, the very awareness that advocacy is an important skill set is generally news to people in healthcare.

Scott: Which blind spot of course benefits that system.

David: Of course. Which makes it doubly important to get this message out: "One skill set you want to have is an advocacy skill set." And what are core advocacy skills? Identifying the gatekeepers and power brokers in your institution. Researching official channels for changing policies and procedure. Discovering informal loops of communication. Finding allies among your peers and in your upline.

But the first step, the hardest step, is to become fully convinced that advocacy is possible. That you should have a voice. That change is possible, even if only incremental at first. That no matter what your pay grade is, you know stuff no one else knows. And that that knowledge is valuable. For out of such grassroots knowledge can come protocols and policies that reduce moral distress, improve patient care, upgrade work conditions, improve lives—and help prevent burnout for everyone. Given that, step up and take a hand in making change!

Scott: So, a crucial thing to ask as one wrestles with moral distress and burnout would be, "Does my system, my institution encourage me to speak? Does my system encourage me to listen carefully and pay attention?" And if not, it's time to engage that advocacy skill set.

In fairness to the human condition, we should acknowledge that CEOs may be as prone to burnout as ICU nurses. But one group is systematically encouraged not to speak much at all, and the other to speak all the time.

David: Think of a hospital unit. You have 30 bedside nurses. Fifteen of them see that many distressing things are going on in the unit. But because the attending physicians and nurse managers above them in the institutional hierarchy are quite often not open to discussion of such concerns, nurses' talk about them commonly turns to weary grumbling and quiet outrage. My question is, What does it take for "grumbling" to become "speaking out"? What does it take for outrage to become action? What must be done to implement the observation and study of the workplace needed for speaking out to be based not on rumor and anecdote, but on factual material that will engage those in power? To some degree, we all have responsibility for the immediate work environment in our institution and we can generate opportunities to improve it.

Scott: Listening and speaking are so fundamental—and potentially massively transformative for systems. I don't mean to say that suddenly everything's going to be better if there's listening and speaking. But institutions are changed by small shifts; and through such shifts institutions can make big changes over time.

David: In just this context it's critical once more to emphasize that burnout does not equal failure. As we have been saying, our institutions are quite often toxic and themselves responsible for burnout. This clearly has nothing to do with personal failure or the lack of personal effort. It is social failure, and on a grand scale.

Yet we must strike a balance, because we know that the institution is going to change very slowly, if at all. We must be able in the meantime to deal with the negativity our institutions bring with them, if we're going to be able to hold a healing space for our patients and hold on to our own healing capacities. At the very same time, we must develop the skills needed for effectively addressing institutions and power structures in their own language.

Scott: One step toward such growth is understanding and accepting that I am a part of what's happening in my institution—in this hospital, in this clinic, on this unit. And because of that, I have influence, however small it may seem. Finding ways to speak out, finding ways to shift my role right where I am has the potential to initiate systemic change.

David: An institution may be viewed as a collection of conversations. If we're in an institution, we're constantly engaged in conversation. Any one of those conversations can be a location for advocacy. It may be changing the tone of our interactions with our peers. It may be speaking to our managers in a different way. It may be going to the diversity office to say my boss is a real problem. It may be going to the Ethics Committee to propose new policies or procedures.

True, if the institution is a series of conversations, some of those conversations are more encrusted than others, some of them have more sediment around them, and some are more influential than others. Granted, all that is true. Yet, the institution is still a series of conversations. We can take that idea and the intention of advocacy into any of those conversations, at any level. And we know by now that an important piece of addressing our own suffering is going to be finding ways to shape those conversations we are already perforce involved in. And to do this effectively, most of us will need to develop and finally master the skills of advocacy.

Dialogue 4
The Field of Suffering Is the Field of Healing

Scott Neely: I want to talk more about how the field of suffering is also the field of healing. I wonder if you could start very basically and say what you mean when you say "the field of suffering." Why speak of it as a field?

David Schenck: In a simple way, one of the first things that comes to mind is a literal field. Battlefield pictures from the stationary fronts of World War I. And, too, the fields that make up Arlington National Cemetery.

And there's physics—magnetic fields and quantum fields.

Yet another element is drawing on Buddhism's understanding of the world. The idea is that everything that is, is interconnected. The doctrine of codependent origination entails that everything is intertwined. Everything is multiply conditioned. Because of that multiplicity, very little is in our control. Everything is always in motion, constantly changing into something else.

Life is painful. And on top of the pain, there is suffering because everything is constantly changing, and this disrupts our mental pictures and our bodily sense of what our lives ought to be. One way to approach that network of interconnectedness is to think of it as a charged field, with distinctive patterns and waves.

Another turn in the teaching is that this field of interconnection is also the field of compassion. We are totally

Into the Field of Suffering. David Schenck with Scott Neely, Oxford University Press.
 DOI: 10.1093/oso/9780197666739.003.0010

interconnected with everything that goes on in the universe. The universe has its own ideas about what it's going to do that are not our ideas—and that leads to suffering. At the same time, we are interconnected in such a way that we feel empathy and kinship and compassion for others. We can be present to them and participate in their healing.

In sum, the imagery of field for me has most deeply to do with interbeing and interconnection. But the first picture—the one that comes immediately to mind—is that of the broad, green swathes of Arlington Cemetery.

Scott: It's a very powerful image. When I hear the word "field," even in the context of suffering, it seems spacious—like a place in which we find ourselves. It has the power to evoke as well not just battlefields in their grim depiction in documentary photographs, but also in their elegiac beauty when they become national parks and open spaces of grassland.

David: Fields of flowers.

Scott: That openness seems, if not to promise, at least to suggest that while there is suffering here, that is not all there is. In this way, it's a very hopeful image.

David: I agree. The image of field suggests possibility of transformation, the possibility of malleability. There is promise that the field itself can be shaped, and one's life along with it.

Whether that's an actual field—battlefield, field of wheat, field of flowers—or a field in physics, what we are talking about is a territory of potentiality. The potentiality that is involved in a life that is constantly changing, and a life that involves multiple dependencies and interconnections. A field that provides for the possibility of great suffering and of great compassion. It's an apposite and powerful image.

Scott: I hear in this idea of the field of suffering that there is woundedness and pain that's happening, but also a greater fertility, a greater productivity, a greater fruitfulness that's possible, there in that same field. In a sense, the field is a greater entity than the suffering itself. It's the sight or the ground of the suffering, but it's not defined by the suffering.

DAVID: Right. But there is an important and difficult point to be made here. Sometimes people are inclined to say that because our suffering is going to be taken up into something larger, we can be sure it will make sense. That's something that I'm inclined to resist.

In my view, it's not that a balance of suffering and healing is guaranteed by the order of things. And when or if we are looking for that balance, or hoping for something like it, or even expecting a balance to show up, we set ourselves up for discouragement, when or if we don't see that happening.

On the other hand, if we say there is this field of possibility, which is constantly manifesting, and that part of its manifestation is suffering, and part of its manifestation is compassion, we can gradually learn how to stand in that field in such a way that we know that *we are that field ourselves*. We are not only in it; we are of it. We're not separate from it. We're a manifestation of it. *We are the suffering and we are the compassion*. This is the very deep place that standing with people who are suffering can take us.

SCOTT: It seems to me, too, that there's a physicality to this image of field. This is in our life, in our body, in this world. This isn't something ephemeral or abstracted or theologized. This is the experience of our lives. We're talking about what it feels like to be in a body and to be in the world.

DAVID: Right. Another way into this whole question of field of suffering and field of healing is the body itself, which is the site of suffering and the site of healing. The body itself is that field of possibility that we are.

The body is constantly changing. The body is constantly being challenged. The body is constantly being nourished. Bodies struggle for balance, for homeostasis. All our suffering is bodily: our spiritual suffering, our mental suffering, our physical suffering. And all our healing is bodily. Not only the healing of the broken bone, or getting over the flu, but the experiences of joy and wholeness and beauty that bring larger healing in our life—all this is fully bodied experience.

Scott: Talk more, please, about suffering and healing together. It's a challenging idea.

David: Let's go at this very concretely. Imagine you are in the presence of a sick person and have entered the field of suffering surrounding them. There is pain, for example, that is projected by their body. Watching somebody who is in pain is a very distinct experience. The person who is in agony projects that agony around them. They project a field of suffering around them.

How does that field become a field of healing? One way this happens is that the patient's field of suffering announces his or her vulnerability, which vulnerability itself invites others to step forward and help. The invitation emanates from the field of suffering, to be picked up by one prepared to offer compassion, to offer assistance, to offer service. People with this vocation respond with the intent to facilitate healing.

The field of suffering is the place where the body and the self and our communities come together to make efforts to restore wholeness. If there is going to be healing, it's going to be something that appears where there is need, where things have been thrown out of order. Where there is deep interconnectedness, there you find a territory of potential, human potential—which is going to result both in suffering and in compassion.

Scott: Yes, I follow that. But we should reiterate the point that, while the field of suffering may have the potential to become the field of healing, there's nothing inevitable about that movement. There's a complexity in how we access the potential for healing, and act on it more or less skillfully. Nothing is certain or guaranteed about any progression from suffering to healing.

David: Yes, a great point. Go back to the physics of the interaction of two fields. There is the person who's in the bed, suffering. There's the field of pain around them. There's a person who enters that space, deliberately walks into the field of suffering, but carrying with them—through their training, through disposition, through intention—a field of healing. They bring with them manifestation of the kinds of energies that can allow a transformation to take place—almost like a flip of polarities, if you will—in the field of suffering that the sick person is inhabiting.

You have this collaborative moment where vulnerability, on the one hand, and capability and capacity, on the other hand, interact. This is the profound moment we may speak of as the crucible of healing. Yet, when those two energetics meet, there is, as you are saying, no guarantee that healing will take place on any level. But whatever potential for healing there is, it is in that place where those fields meet. Together they become a larger field that will move and morph in one direction or another.

Dialogue 5
The Fulcrum and the Great Compassion

SCOTT NEELY: Fundamentally, what I hear you saying is that in our ordinary limitations—the absolute limitations of our lives—*in these very limitations there is capacity*. There is something there for us to nurture and to develop.

I hear something that honors suffering as it is, as an experience—and honors it for its potential for developing capacity. Which potentiality doesn't deny or obfuscate the harsh reality of suffering. If we can hold this paradox, there's a fulcrum here. There's a power, a leverage point—very difficult to maintain, but immense.

DAVID SCHENCK: Let's go with this terrific image: What is that fulcrum? And, next, how does one hold to the fulcrum?

First off, the literal image. In one view, the fulcrum point provides leverage; it amplifies force. In another view, the fulcrum is the balance point—the point that allows the two opposite ends of a lever to function together as one, despite moving in opposite directions.

Now call to mind the complicated dance of healing initiated by the vulnerability of the patient and the healer's response. This is a dance of gift, capacity, and woundedness, which in its most developed forms can strike a point where one person is together with another, embracing the enormous range of human vulnerability and distress, and compassion and presence, all in one joint movement. That to me is a fundamental definition of healing: the experience of encompassing all that in an encounter

Into the Field of Suffering. David Schenck with Scott Neely, Oxford University Press.
 DOI: 10.1093/oso/9780197666739.003.0011

with another. And that point of balance, at the intersection of all those vectors: *that's the fulcrum.*

Scott: We've spoken together about the Great Compassion [Dalai Lama 2020]. That would be one way to describe that place, that fulcrum. It makes room for us and offers solace and peace—the phrase "the Great Compassion" does that. But it is also activating. This is what is in us. This is what is expected of us. This is what we can offer. And if we don't choose it, it's less apparent in the world, and less apparent in us. We experience it less, which is a great motivation to choose it!

David: One of the extraordinary privileges granted to people who choose to work with others in extreme suffering is the invitation to be with them at this fulcrum point. We are there with them looking over the edge of the world, as I described it earlier, where the whole range of human experience can be grasped. We're there in a moment of potentially enormous directness, enormous honesty, enormous giving. My hospice nurse friend Priscilla used to say, "I do hospice work because there's no time for bullshit. You have to be fully there; and you have to speak the truth." Almost everyone I know who cares for people who are in extremity will recognize those moments where, as one doctor described it, "The whole world falls away. It's you and the patient there together." There are those moments where time itself is subsumed in the intensity of connection.

We stand with those who suffer in part for them, of course. But also, in part, to increase our own vulnerability and our own suffering.

Scott: Wait, to increase our own suffering?!

David: Weird, right? But, yes, to increase our vulnerability to suffering, to open ourselves more, so we can learn more from it and about it. To learn more, and to grow, by going into suffering and being opened by it. It is a paradox, and a source of authentic power for service. After all, illness and suffering are inevitable in being a being, and being human. We all experience wounding throughout our lives. Fate and karma are universally wounding.

Scott: I understand entering the field of suffering to help. But I had not reckoned with the thought of entering the field to be opened by suffering ourselves.

David: Coming to terms with one's own woundedness entails incorporating deeply into oneself and one's body the truth about the field of suffering being the field of healing. Woundedness is vulnerability, isolation, pain—all of which are aspects of interconnection and interbeing. Healing is growth, sharing, capacity—all of which likewise are aspects of interconnection and interbeing. The ongoing challenge and opening in each of the six phases of the healing vocation is how woundedness is held: curing wounds, fixing them; being crippled by them, taught by them, fed by them. Which, in turn, at its deeper level, remains an ongoing question about the relationship of healing and suffering: Are the two balanced? Which pole dominates? How can both be held? And all this is taken up in twin themes of "the fulcrum" and "the Great Compassion."

Scott: Yet it seems to me there is a danger here of becoming too self-focused, losing the partnership, losing the mutuality. As if I am here ultimately for my own personal or spiritual development. There is more purpose to what you are suggesting than these smaller views allow.

David: As one grows into the vocation, there can be an ever-increasing sense of what the compassionate heart demands—and of the resources the compassionate heart offers. One way to think about what breakthrough involves is seeing it, and the depletion that prepares the way for it, as a painful but ultimately rewarding process of having our capacity expanded. It hurts, to be sure. But we are being stretched so we can hold more, so we can stay compassionate and balanced and present in the face of worse and worse suffering. We become able to be present to situations that call up horrible memories out of our past. It becomes possible to be consistently a still collecting point, even as conflict and distress and even agony move around us. When we have not walled off pieces of our lives that are challenging or painful, they become part of what is present when we are present. They are present in us. There's potential for a whole different level and range of connection that wouldn't be there without them.

Scott: Learning to be present through our own pain, to be more compassionate with the pain of others.

David: Right. And here we can begin to glimpse something truly terrifying: the possibility of the ego being commandeered in service of the heart center. As the cycles of depletion and breakthrough repeat, more and more of one's ego, one's self, one's life is commandeered by the heart. Eventually, this progression may feel like the darkest night of all, the deepest depletion, the most total burnout. But it is the opening for the greatest realization; it is full of potential. True, we learn little or nothing about this level of service in professional literature. And that is appropriate. It is only from the saints and the mystics that we can get this guidance. If there is a mystery in human life equal to that of suffering, it is that of the mystery of compassion.

APPENDIX

A Practice Calendar

The following practice calendar is designed to assist the reader in developing key skills discussed in Chapters 4 and 5, and Dialogue 3, into practices that inform daily life and work. The practice calendar (1) shows the incremental and developmental relationships among the exercises and (2) explains how to work with them concretely in specific work environments.

Week 1

Pay attention to which areas in your body carry the most tension. Work with the physical pain. Can you find its epicenter? Does your awareness, your non-judgmental awareness, change your experience of it?

Week 2

Pay attention to times of day your body is most tense. Pay attention to the spaces and settings where this occurs. By midweek, whenever and wherever you are feeling challenged, practice straightening your spine, planting your feet, exhaling thoroughly, and taking deep belly breaths.

Week 3

Once more, pay attention to times of day your body is most tense. Pay attention to the spaces and settings where this occurs. This week, at those same most stressful times and places, do the pause–center–shift practice.

Week 4

This week look for mirroring behaviors you commonly engage in. Do a typology: Which are most common in which situations? Which are most common with which people or types of people: physicians, nurses, residents, students; administrators, counselors, chaplains, technicians, social workers; patients, parents of patients, partners of patients, children of patients.

Develop and practice ways to disrupt the mirroring without disrupting your connection to the people with whom you're interacting.

Week 5

This week, focus on hot-button people and situations. Begin with the mildly annoying. By the end of the week, move on to the more challenging; but don't yet take on the most challenging people in your life. Ask yourself these questions: Are the hot spots mostly in one place in your body (e.g., neck, shoulders, lumbar spine, belly, solar plexus, throat)? What brings most relief? Breathing into them? Stretching? Walking it off? Leaving the unit? Leaving the building? Talking? Journaling? Screaming?

Give yourself permission not to be trapped: either in a specific physical location or in a state of physical discomfort in your body.

Week 6

This week, with those same people and situations, do the pause–center–shift practice. Move on to try these practices with your most challenging colleagues and patients.

Week 7

Activism opportunities. Review the discussion of advocacy in Dialogue 3. This week think like a community organizer. Canvas your peers. Identify workplace issues and institutional resources. Be guided by your body and its response to stress, as much as by your mind, in deciding on action priorities.

Week 8

Focus on the gratitude practice this week. It's described in detail in Chapter 5. Be grateful for your colleagues and for your patients.

Follow-Up

Repeat the eight-week cycle. Continue the cycle for one year. Or develop your own calendar.

Meanwhile, to support these focused efforts, the development of a mindfulness practice is recommended. People with medical training generally respond well to the approach taken for training Google engineers found in Chade-Meng Tan's *Search Inside Yourself*, or the approach developed in the context of pain management and medical centers presented in *Full-Catastrophe Living* by Jon Kabat-Zinn. Jack Kornfield is a clinical psychologist who also trained as a Buddhist monk. This combination helps make his teaching in his many works, such as *Meditation for Beginners*, accessible for healthcare professionals. Also useful is Christiane Wolf and Greg Serpa's *A Clinician's Guide to Teaching Mindfulness* (Tan 2012; Kabat-Zinn 2013; Kornfield 2008; Wolf and Serpa 2015).

Another deeply rewarding, supportive practice is offering loving-kindness to all people, including those who benefit you most, and those you find most challenging. Sharon Salzberg's classic, *Lovingkindness: The Revolutionary Art of Happiness*, provides an easily accessible approach to this form of meditation (Salzberg 2002).

References

Camus, Albert. 1991. *The Plague*. Translated by Stuart Gilbert. New York: Vintage.

Chapple, Helen S. 2010. *No Place for Dying: Hospitals and the Ideology of Rescue*. Walnut Creek, CA: Left Coast Press.

Chapple, Helen S. 2015. "Rescue: Faith in the Unlimited Future." *Society* 52 (5): 424–29.

Churchill, Larry R., Joseph B. Fanning, and David Schenck. 2013. *What Patients Teach: The Everyday Ethics of Health Care*. New York: Oxford University Press.

Churchill, Larry R., and David Schenck. 2008. "Healing Skills for Medical Practice." *Annals of Internal Medicine* 149 (10): 720–24.

Dalai Lama [Tenzin Gyatso] and Thubten Chodron. 2020. *In Praise of Great Compassion*. Somerville, MA: Wisdom Publications.

Eliade, Mircea. 2004. *Shamanism: Archaic Techniques of Ecstasy*. Translated by Willard R. Trask. Princeton: Princeton University Press.

Epstein, Elizabeth G., and Ann B. Hamric. 2009. "Moral Distress, Moral Residue, and the Crescendo Effect." *Journal of Clinical Ethics* 20 (4): 330–42.

Epstein, Elizabeth G., Phyllis B. Whitehead, Chuleeporn Prompahakul, Leroy R. Thacker, and Ann B. Hamric. 2019. "Enhancing Understanding of Moral Distress: The Measure of Moral Distress for Health Care Professionals." *AJOB Empirical Bioethics* 10 (2): 113–24.

Frank, Arthur W. 2002. *At the Will of the Body: Reflections on Illness*. Boston: Houghton Mifflin.

Frank, Arthur W. 2013. *The Wounded Storyteller: Body, Illness and Ethics*. Chicago: University of Chicago Press.

Gallese, Vittorio. 2003. "The Roots of Empathy: The Shared Manifold Hypothesis and the Neural Basis of Intersubjectivity." *Psychopathology* 36 (4): 171–80.

Halifax, Joan. 2018. *Standing at the Edge: Finding Freedom Where Fear and Courage Meet*. New York: Flatiron.

Hamric, Ann B., Christopher T. Borchers, and Elizabeth G. Epstein. 2012. "Development and Testing of an Instrument to Measure Moral Distress in Healthcare Professionals." *AJOB Primary Research* 3 (2): 1–9.

Hatfield, Elaine, John T. Cacioppo, and Richard L. Rapson. 1994. *Emotional Contagion*. Cambridge: Cambridge University Press.

Huber, Cheri. 1999. *The Key: And the Name of the Key Is Willingness*. Rev. ed. Designed and illustrated by June Shiver. Mountain View, CA: Keep It Simple Books.

Huber, Cheri. 2000. *How to Get from Where You Are to Where You Want to Be*. Carlsbad, CA: Hay House.

Huber, Cheri. 2018. *How You Do Anything Is How You Do Everything*. Rev. ed. Edited and illustrated by June Shiver. Mountain View, CA: Keep It Simple Books.

Jameton, Andrew. 2017. "What Moral Distress in Nursing History Could Suggest About the Future of Health Care." *AMA Journal of Ethics* 19 (6): 617–28.

John of the Cross. 1987. *Selected Writings*, edited by Kieran Kavanaugh. New York: Paulist Press.

Johnson, Robert A. 1995. *The Fisher King and the Handless Maiden: Understanding the Wounded Feeling Function in Masculine and Feminine Psychology*. New York: HarperCollins.

Johnson, Will. 2000. *Aligned, Relaxed, Resilient: The Physical Foundations of Mindfulness*. Boston: Shambhala.

Joy, W. Brugh. 1990. *Avalanche: Heretical Reflections on the Dark and the Light*. New York: Ballantine.

Joy, W. Brugh. 2008. "Oral Teaching." Heart Beat Workshop, October 27–November 3, Lone Pine, CA.

Joy, W. Brugh. 2010. "Healing and the Unconscious" (excerpt). ThinkingAllowedTV. Thinking Allowed video, with Jeffrey Mishlove, August 20, 2010. https://www.youtube.com/watch?v=BhQWMSANHdo.

Kabat-Zinn, Jon. 2013. *Full-Catastrophe Living: Using the Wisdom of Your Body and Mind to Face Stress, Pain, and Illness*. Rev. ed. New York: Bantam.

Katz, Renee S., and Therese A. Johnson, eds. 2006. *When Professionals Weep: Emotional and Countertransference Responses in End-of-Life Care*. New York: Routledge.

Kaufman, Sharon R. 2015. *Ordinary Medicine: Extraordinary Treatments, Longer Lives, and Where to Draw the Line*. Durham, NC: Duke University Press.

Kornfield, Jack. 2008. *Meditation for Beginners*. Boulder, CO: Sounds True.

Levine, Peter A. 1997. *Waking the Tiger: Healing Trauma*. With Ann Frederick. Berkeley, CA: North Atlantic Books.

Levine, Peter A. 2010. *In an Unspoken Voice: How the Body Releases Trauma and Restores Goodness*. Berkeley, CA: North Atlantic Books.

Nouwen, Henri J. M. 1979. *The Wounded Healer: Ministry in Contemporary Society*. New York: Doubleday Image.

Ray, Reginald. 2016. *The Awakening Body: Somatic Meditation for Discovering Our Deepest Life*. Boulder, CO: Shambhala.

Remen, Rachel Naomi. 1999. "Helping, Fixing or Serving?" *Shambhala Sun*. September 1999. https://www.mentalhealthsf.org/wp-content/uploads/2020/01/HelpingFixingServing-by-Rachel-Remen.pdf.

Rodney, Patricia A. 2017. "What We Know About Moral Distress." *American Journal of Nursing* 117 (2 Suppl 1): S7–S10.

Rotenstein, Lisa S., Matthew Torre, Marco A. Ramos, Rachael C. Rosales, Constance Guille, Srijan Sen, and Douglas A. Mata. 2018. "Prevalence of Burnout Among Physicians: A Systematic Review." *JAMA* 320 (11): 1131–50.

Rothschild, Babette. 2006. *Help for the Helper: The Psychophysiology of Compassion Fatigue and Vicarious Trauma*. With Marjorie L. Rand. New York: Norton.

Rushton, Cynda Hylton. 2009. "Ethical Discernment and Action: The Art of Pause." *AACN Advanced Critical Care* 20 (1): 108–11.

Rushton, Cynda Hylton, ed. 2018. *Moral Resilience: Transforming Moral Suffering in Healthcare*. New York: Oxford University Press.

Salzberg, Sharon. 2002. *Lovingkindness: The Revolutionary Art of Happiness*. Boston: Shambhala.

Schenck, David. 2015. "Death's Broker: The Ethics Consultant in the ICU." *Society* 52 (5): 448–61.

Schenck, David, and Larry R. Churchill. 2011. *Healers: Extraordinary Clinicians at Work*. New York: Oxford University Press.

Sobel, Richard. 2008. "Beyond Empathy." *Perspectives in Biology and Medicine* 51 (3): 471–78.

Steindl-Rast, David. 2017. "A Grateful Day." Network for Grateful Living. YouTube video, August 22, 2017. https://www.youtube.com/watch?v=zSt7k_q_qRU.

Tan, Chade-Meng. 2012. *Search Inside Yourself: The Unexpected Path to Achieving Success, Happiness (and World Peace)*. New York: HarperOne.

Ulrich, Connie M., and Christine Grady, eds. 2018. *Moral Distress in the Health Professions*. Cham, Switzerland: Springer.

Van der Kolk, Bessel A. 2015. *The Body Keeps the Score: Brain, Mind, and Body in the Healing of Trauma*. New York: Penguin.

Vercio, Chard, Lawrence K. Loo, Morgan Green, Daniel I. Kim, and Gary L. Beck Dallaghan. 2021. "Shifting Focus from Burnout and Wellness Toward Individual and Organizational Resilience." *Teaching and Learning in Medicine* 33 (5): 568–76.

Wolf, Christiane, and J. Greg Serpa. 2015. *A Clinician's Guide to Teaching Mindfulness: The Comprehensive Session-by-Session Program for Mental Health Professionals and Health Care Providers*. Oakland, CA: New Harbinger.

Index

For the benefit of digital users, indexed terms that span two pages (e.g., 52–53) may, on occasion, appear on only one of those pages.

Tables are indicated by *t* following the page number